Environmental and Food Virology
Impacts and Challenges in One Health Approach

Editors

Gislaine Fongaro

Laboratory of Applied Virology
Federal University of Santa Catarina
Department of Microbiology, Immunology and Parasitology
Santa Catarina, Brazil

David Rodríguez-Lázaro

Microbiology Division, Department of Biotechnology
and Food Science, Faculty of Sciences
University of Burgos
Burgos, Spain

Doris Sobral Marques Souza

Laboratory of Applied Virology
Federal University of Santa Catarina
Department of Microbiology, Immunology and Parasitology
Santa Catarina, Brazil

CRC Press
Taylor & Francis Group
Boca Raton London New York

CRC Press is an imprint of the
Taylor & Francis Group, an **informa** business

A SCIENCE PUBLISHERS BOOK

Cover credit: Ksandrphoto - Freepik.com

First edition published 2023
by CRC Press
6000 Broken Sound Parkway NW, Suite 300, Boca Raton, FL 33487-2742

and by CRC Press
4 Park Square, Milton Park, Abingdon, Oxon, OX14 4RN

CRC Press is an imprint of Taylor & Francis Group, LLC

Library of Congress Cataloging-in-Publication Data (applied for)

ISBN: 978-1-032-20419-2 (hbk)
ISBN: 978-1-032-20421-5 (pbk)
ISBN: 978-1-003-26349-4 (ebk)

DOI: 10.1201/9781003263494

Typeset in Times New Roman
by Radiant Productions

Preface

||||||||||||||||||||||||||||||||||||||||||||||||||||||||||||||||||||||||||||||||||||||||||||||||||||||||||||||||||||||||||||||||||||||||||||||||

The present book "Environmental and Food Virology: Impacts and Challenges in One Health Approach" organized by the respected scientists Gislaine Fongaro, David Rodríguez-Lázaro and Doris Sobral Marques Souza, presents the most important issues regarding the main viruses that may be potentially present in the aquatic environment and foods as well as their direct or indirect relationship with diseases, nutrition and world economy, perfectly integrating these issues on the homeostasis of the One Health concept. The current moment in which our planet lives facing pandemic and epidemic diseases caused by viruses, draws more and more attention to the pathogenic viruses present in the environment and in food and how important it is to monitor, quantify, choose representative biomarkers and discover new viruses. This book aims to shed new light on these important concepts, highlighting their importance for epidemiological issues, analysis of real risks of contamination as well as the use of viruses as biotechnological tools to find an expected balance of the One Health concept aiming to attain good health for all the living beings on our planet.

The authors dedicate this book to Dr. Célia Regina Monte Barardi, who started the line of research in Environmental Virology at the Federal University of Santa Catarina and contributed a lot to the training of human resources and virology around the world.

Contents

||||||||||||||||||||||||||||||||||||||||||||||||||||||||||||||||||||||||||||||||||||||||||||||||||||||||||||||||||||||||||||||||||||||||||||||||||||||||||||||||||||||||||||||||

Chapter 1

An Overview of Viruses Transmitted by Water and Food

Main Actors and the Main Transmission Routes

Aline Viancelli,[1,*] *Bruna Petry*[2] and *William Michelon*[1]

IIIIIIIIIIIIIIIIIIIIIIIIIIIIIIIIIIIIIIIIIIIIIIIIIIIIIIIIIIIIIIIIIIIIIIIIIIIIIIIIIIIIIIIIIIIIIIIIIIIIIIIIIIIIIIIIIIIIIIIIIIIIIIIIIIIIIIIIIIIIIIIIIIII

1. Introduction

1.1 How did viruses' discovery change the way we see the environment?

More than a century ago, in 1898, a scientist called Martinus Beijerinck described *contagium vivum fluidum,* a filterable substance capable of starting the tobacco mosaic disease in living leaves, and he was the first to call this liquid a 'virus' (Lecoq 2001). Many years later, this virus was characterized as the *Tobacco mosaic virus* (Scholthof 2004). Beijerinck also suggested that this agent should enter a live cell for a successful multiplication process, and gave the first steps to differentiate viruses from bacteria (Letarov 2020). Ever since, a large number of viruses have been reported all around the world, integrating all the knowledge in the research field of virology. Virology is a research field that led to significant changes in the scientific world since its beginning. Within the virology studies, there is a special division called "environmental virology" which was one of the first areas to look at nature in a holistic view, considering the interactions among humans, animals, and the environment.

---

[1] Universidade do Contestado, Programa de Mestrado Profissional em Engenharia Civil, Sanitária e Ambiental. Concórdia – Santa Catarina, Brazil.
[2] Escola Superior de Agricultura "Luiz de Queiroz", Universidade de São Paulo – São Paulo, Brazil.
* Corresponding author: alineviancelli@unc.br

Environmental virology began when Joseph Melnick (Melnick 1947) realized that the poliovirus (PV) strains present in sewage were spreading through environmental routes, and alerted the scientific community about the importance of environmental research for virus monitoring. Later, in the 1950s, New Delhi (India) authorities and scientists reported a huge waterborne outbreak of hepatitis associated with the rupture of a wastewater treatment plant and the flow of wastewater into the river, which was used as drinking water supply (Chuttani et al. 1966). Since these events, a special group of viruses have been associated with fecal-oral transmission routes. The group was named enteric viruses due to their replication cycle that is performed on the enterocytes cell present in the intestine (Kolawole and Wobus 2020). This group has been studied in food and environmental samples, such as water, sediments, shellfish and a large number of vegetables (Upfold et al. 2021). Nowadays, several viruses' families have been reported to be present in these matrices. All these viruses that can infect human cells (or other animals), have the potential to be pathogenic, depending on the host health or the virus cell entry mechanisms of evolution. As a result, they have been studied more and more.

Since the presence of viruses was detected in the environment, researchers have worked on the implementation and improvement of sample concentration methods (Ikner et al. 2012), to evaluate the virus particles present in different matrices, such as water, wastewater, seawater, oyster, and sediments (Hernandez-Morga et al. 2009, Metcalf et al. 1995, Stals et al. 2011). These first techniques used to identify viruses in environmental samples had disadvantages such as low volume of sample capacity or the simultaneous concentration of toxic substances that damaged the cell culture (Ikner et al. 2012).

Later, the membrane filtration methods were included in the concentration protocols of samples, adsorbents were tested, flocculation with skimmed milk, polyethylene glycol (PEG), and so on. However, none of them was perfect to concentrate all the viruses present in the sample. Electronegative or electropositive membranes, and those techniques that use any type of membranes, are not suitable to be applied to wastewater, sludge, raw manure or sediment samples, because of the huge amount of organic matter present in these matrices (Masclaux et al. 2013). So, the choice of virus concentration methodology to be applied will depend on the sample's characteristics (Corpuz et al. 2020).

Together with the sample's concentration methods, the virus' detection alternatives were also changed over time. First, the cell culture procedures were used to evaluate the virus infectivity capacity. However, cell culture was labor intensive and some viruses were fastidious to propagate, or did not have any permissive cell lineage. This gap was filled by the molecular tools, such as polymerase chain reaction (PCR) (Adefisoye et al. 2016) that in the beginning was disposable as a qualitative method, and years later as a quantitative technique with the qPCR development (Aw and Rose 2012). Integrated Cell Culture and PCR (ICC-PCR) have also been used with the aim to evaluate the infectivity of viruses that are not able to form plaque lyse (Girones and Bofill-Mas 2013).

Genome sequencing techniques, such as next generation sequencing, improved the classification of the isolated strains, allowed metagenomic studies of microorganisms present in a sample and helped in clarifying the spread of specific strains and occurrence of outbreaks (Bibby et al. 2011). Microscopy advances helped researchers to understand the virus particles' structures (Richert-Pöggeler et al. 2019). The cell culture continues to be an excellent alternative to understand the virus infection behavior, when associated with application of green fluorescent protein (GFP) biomarker (Galbraith et al. 1998, Wieczorek et al. 2020). Finally, the molecular beacon techniques have been applied to understand the virus entry on cells in real time observation (Ganguli et al. 2011, Sukla et al. 2021).

Despite all this investigation/monitoring advances, the waterborne diseases are still a public health problem all around the world, especially for children below 5 years of age living in developing countries. In low income countries, diarrhea was the eighth cause of death in 2019 (WHO 2022). However, it is important to highlight that subclinical or asymptomatic infections, or absence of confirmation diagnostics still influences epidemiological studies (Miagostovich and Vieira 2017). Because of this, the United Nations established the 17 Sustainable Development Goals by 2030, where goal number 6 refers to improvement of clean water and sanitation for all, and the need of more studies in this issue (United Nations 2020). Other important support to researchers in the virology field was the formation of International Society for Food and Environmental Virology, where scientists can improve the knowledge about environmental transmission routes of viruses and share experiences (Miagostovich and Vieira 2017).

Another advancement in the research field of environmental virology was the creation of the "One Health" concept. In the beginning of 21st century, the zoonotic viruses were emerging and re-emerging, such as bird flu, the Ebola and Zika viral epidemics, due to the interdependent relationship of human health, animal health, and ecosystem health (Destoumieux-Garzón et al. 2018). In a huge effort, governments and scientists all around the world realized that an interdisciplinary collaboration was needed to prevent the spread of viruses and possibilities of pandemics (Gibbs 2014).

The "One Health" concept was proposed as a global strategy aiming to integrate professionals from different areas to deal with diseases occurrence, spread and control of pathogens, in a holistic perspective (Zinsstag et al. 2015). The concept was adopted by the three most important agencies related to health: Food and Agriculture Organization (FAO), World Health Organization (WHO) and World Organization for Animal Health (OIE) (Gibbs 2014). The environmental virology became so important to One Health purposes, that the term "virosphere" has been used to describe the space where viruses are found and the space they influence, highlighting the importance of these viruses for ecological interactions (Richert-Pöggeler et al. 2019).

Considering that all these conceptual and technological changes occurred throughout history, scientists made elementary questions: how do these harmful viruses reach the environment and which are the hotspots to prevent dissemination and disease outbreaks?

1.2 How do these viruses reach the environment?

These viruses are excreted in the feces, and could remain viable in raw (untreated) or inadequately treated wastewater, sewage or animal manure (Carducci et al. 2006), especially considering that traditional wastewater treatment processes do not completely reduce viral loads of enteric viruses. In this sense, when the wastewater is disposed in the environment, the pathogens reach the water bodies, such as springs, rivers, or marine estuaries. Once in the water, the pathogens can enter the human body through various exposure possibilities such as intake of contaminated drinking water, direct skin contact or by accidental ingestion of recreational waters, consumption of contaminated food such as vegetables or filter-feeding bivalve shellfish (Rodríguez-Lázaro et al. 2012). When infected, the individuals reinitiate the transmission cycle (Figure 1).

The contamination of water is a consequence, mainly, of the growth of civilizations, construction of cities near rivers and the lack of wastewater treatment. Historically, the water quality was based on the visual observation of "clarity" of the water, without any on the modern chemical and microbial analysis (Hall and Dietrich 2000). One of the oldest registers of a water treatment method was performed by ancient Egyptians (1500 BC) using alum application to remove the solid particles by coagulation (Ahmed et al. 2020). The ancient Hindus also left documents (before Common Era) where they registered the indication for people to boil the water before drinking (Hall and Dietrich 2000).

The water treatment plants similar to the modern systems were put into operation in the 1880s and 1890s, when the first chloride application was performed for microbial removal treatment. Further, in 1942, a drinking water standard including bacteriological sampling was adopted (Hall and Dietrich 2000). However, in 1984 the inefficiency of water treatment systems for removal of enteric viruses was pointed by a study that showed that 83% of the water samples collected along different steps of the treatment were positive to the presence of enteric viruses (Keswick et al. 1984). Ever since, virologists have indicated the need of monitoring these viruses on treated drinking water.

Nowadays there are other forgotten scenarios by the water treatment companies: the rural areas. At these distant places, where water treatment systems do not exist, water used for drinking purposes is collected from springs, wells and groundwater, which have been reported with high contamination loads of viruses (Fongaro et al. 2015, Fornari et al. 2021).

Besides drinking water, recreational water has also been pointed out as a key health risk factor leading to enteric virus contamination primarily through contact during swimming or other water sports (Federigi et al. 2019). The same is observed for outbreaks associated with treated recreational water, such as swimming pools, where in the period of 2015–2019 in the United States at least 208 outbreaks were reported (Hlavsa et al. 2021). The climatic changes are increasing the precipitation volume, and consequently the frequency and intensity of events such as floods, this water directly affects the spread of viruses (Hofstra et al. 2019, Cadamuro et al. 2021).

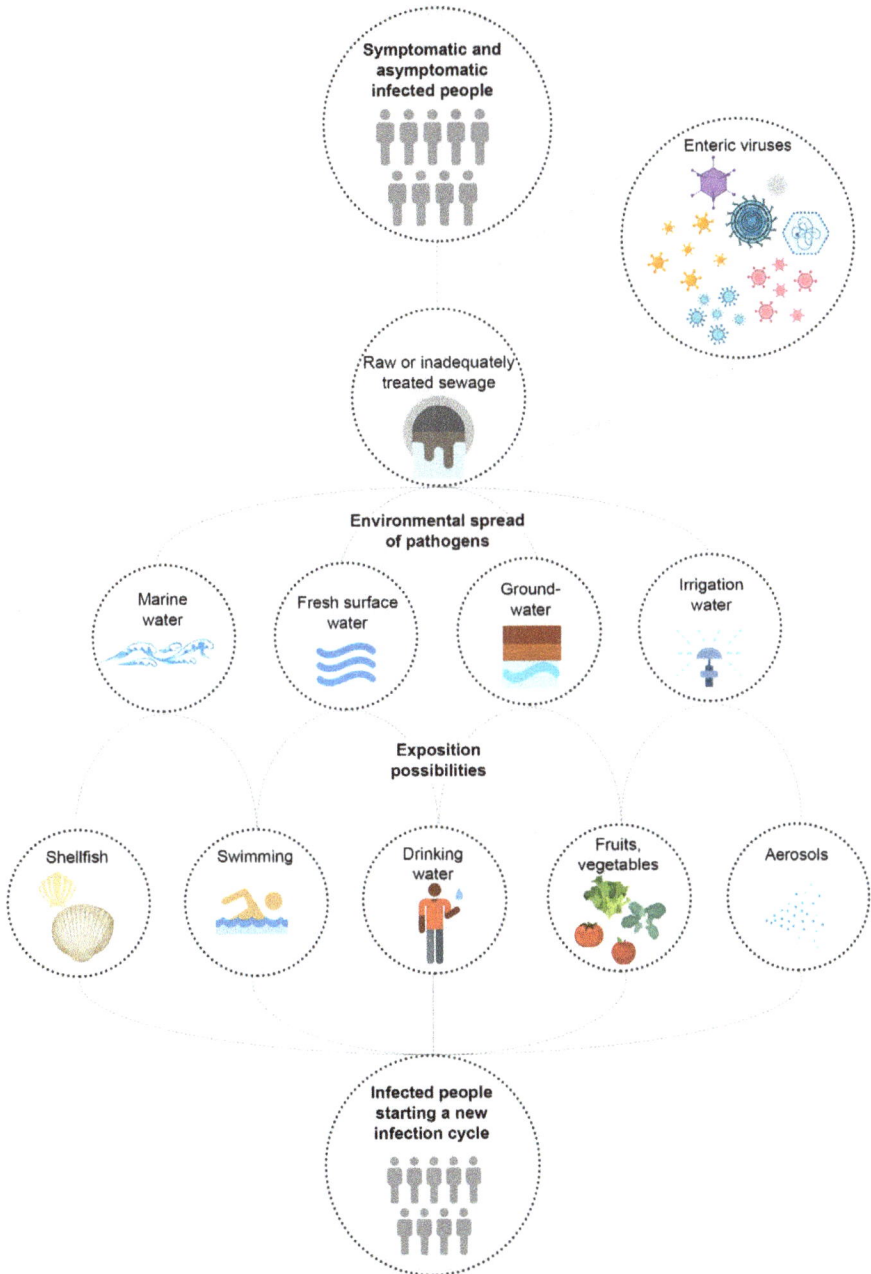

Figure 1. Schematic representation of the possibilities involved in the environmental virology contamination cycle.

If viruses present in wastewater and other water bodies are considered as possible cause of contamination of fruits and vegetables, it could open a new perspective for environmental spread of pathogens. The presence of virions (complete and infective virus particles) on food always represents a health risk, so it is important to evaluate food properties and storage conditions to understand the virus behavior and persistence in these matrices (Roos 2020). Information such as the chemical interaction between the enteric viruses and the food surface, that occurs by ionic and hydrophobic interactions, as van der Waals forces, are important when considering the success of virus removal process applied on the vegetable surfaces (Sánchez and Bosch 2016). In addition, studies have reported that if contaminated vegetables are kept at refrigerator temperature (4°C), the virus survival period is longer than vegetables left at room temperature (Koopmans and Duizer 2004, Kurdziel et al. 2001).

The vegetable association with foodborne illness outbreaks became a new concern with the increase of consumption of raw vegetables or minimally processed (leaves, tubers and roots) in daily diet, linked with the absence of a proper and effective decontamination step along the salads production chain (Warriner 2005). This contamination can occur by irrigation with water containing pathogenic microorganisms, internalization of pathogens by the roots, or during handling steps such as cutting, slicing, skinning and shredding by infected workers during the food preparation processes (Kokkinos et al. 2017). When the internalization occurs, the surface sanitization procedures of postharvest vegetables are not effective, due to the fact that sanitizers do not enter plant cells (Fuzawa et al. 2021).

However, the greater concern about vegetable contamination routes begins at the cultivation step, where the wastewater is generally applied on growing vegetable chains as a biofertilizer (Raina et al. 2020). Besides the recommendation of the Environmental Protection Agency (EPA) for the application of biosolids derived from manure treatment only if the enteric viruses load is below 4 plaque-forming units per gram biosolids (other microorganisms also have to be monitored), the majority of the cultivation practices use untreated wastewater as fertilizer (de Matos Nascimento et al. 2020, Gerba et al. 2017). It is also worth mentioning that the viral human pathogens can be accidentally exposed to irrigation water through heavy rainfall and consequent surface run-off after wastewater application on soil, or illegal discharge of waste on environment (Pepper et al. 2019). Another critical point about the vegetable contamination is that enteric viruses can stay viable for up to four months in subsurface soil layers (Warriner 2005).

Virus contamination has been observed in leafy green vegetables and berry fruit production chains (Kokkinos et al. 2017). One of the most remembered outbreaks related to this issue occurred in 2003, when 1000 illness cases were reported after the consumption of green onions contaminated with *Hepatitis A virus* (HAV) (Wheeler et al. 2005). *Human norovirus* and *Hepatitis A virus* have been reported to be transmitted in fresh and frozen strawberries (Bozkurt et al. 2021, Lévesque et al. 2021).

Viruses' characteristics such as size and surface structure can increase the ability of the virus to internalize through the vegetable root system, with higher internalization in plants cultivated on hydroponic systems than those cultivated on soil (Hirneisen et al. 2012). However, it is important to highlight that the presence of viruses in the hydroponic solution does not guarantee internalization by the vegetables. Wei et al. (2011) evaluated the transpiration rate of lettuce at difference humidity conditions. The authors reported that, at 70% of relative humidity, the lettuce transpiration rate induced a higher internalization of Murine norovirus type 1 virus in the leafs, if compared to the observed in the lettuce leafs cultivated at 99% relative humidity's condition. In a similar study, it was demonstrated that in romaine lettuce cultivated under abiotic stress such as drought, the internalization and dissemination of *Murine norovirus* rates increase (DiCaprio et al. 2015).

Another foodborne route associated with environmental virology is through the consumption of bivalves mollusks (shellfish). The first important highlight about the bivalves is the place where they are cultivated, which is, seawater areas. Seawater pollution has increased in huge proportions over the past decades, due to anthropogenic pressure in coastal areas (Moresco et al. 2012). The second important highlight is about the mollusks' biology. They are filter feeders, that filter large volumes of seawater during their natural feeding and consequently concentrate suspended particles found in the water column, including a variety of human pathogens like bacteria and viruses (Mcleod et al. 2009, Polo et al. 2015).

This filtration capacity has one advantage and one disadvantage. The advantage is that bivalves have been considered as suitable bioindicators of pollution to the marine environment and could be used for monitoring purposes, such as the presence and quantification of pathogenic microorganisms (Souza et al. 2018). It has been reported that the concentration of enteric virus particles on oysters' organs can be 100 times higher than those observed in contaminated seawater (Butt et al. 2004). The disadvantage is that the gastronomy way to consume bivalves mollusks such as oyster is raw (without cooking). Therefore, the presence of pathogens in this food is a health risk to consumers, and outbreaks caused by contaminated shellfish have been reported all around the world (Bellou et al. 2013, Savini et al. 2021).

Some alternatives have been studied aiming to solve or minimize this problem, such as depuration processes, or heat treatment of the bivalves before consumption (Souza et al. 2018). The heat treatment often means to put the oyster in boiling water for at least 90 seconds, however, depending on the mollusk species, this time could not be enough to open the shell and increase the temperature of the organs and killing the pathogens (Sobral Marques Souza et al. 2015, Souza et al. 2018).

One important highlight about enteric viruses in vegetables and other food, is that the global production of this product is increasing every year, and consequently the exportation is also increasing, making these vegetables "travel" to the most distant places in the world, and also the viruses present on them (Balali et al. 2020, Hess and Sutcliffe 2018).

With the help of robust data collection over the years, all these studies have contributed, towards understanding and pointing out the hotspots of enteric virus dissemination and the possible situations related to occurrence of outbreaks. This knowledge will be the guide for future decisions on the One Health approach.

1.3 Who are these viruses?

The biology of enteric viruses comprises a broad range group, composed of more than 150 different species of pathogens (Carducci et al. 2009) that colonize the gastrointestinal tract and, due to this, are excreted in feces, and then found on water or wastewater samples. The most commonly reported viruses are *Aichi virus, Astrovirus, Cosavirus A, Coxsackievirus* B, *Enterovirus* (A–D), *Hepatitis A virus, Hepatitis E virus, Human bocavirus* type 1–4, *Human adenovirus* 40, 41, *Mastadenovirus* (A–F), *Human polyomavirus* (HPyV), *Norovirus* genotype I and II, *Rotavirus A, Sapovirus, Torque teno virus* (Carter 2005, Fong and Lipp 2005, Upfold et al. 2021). All these viruses are excreted on feces in concentration as high as 10^{13} virus/g of feces (so that they persist for long periods in environmental samples) and have low infectious dose (< 20 particles) required to cause illness (Upfold et al. 2021). *Human norovirus* (HuNoV), for example, has an infective dose of 1 to 10 virions (complete virus particles capable of initiating an infection process) (Bosch et al. 2018).

These viruses are non-enveloped particles, and this characteristic makes them naturally highly stable in water environments, particularly when adsorbed to organic matter or sediments (Hejkal et al. 1981, Seitz et al. 2011), and when these materials are revolved and resuspended, the pathogens return to the water column (Sassi et al. 2020). In this condition, the virus particle remains protected from environmental adversities such as sun UV light, temperature oscillation, and so on (Rzeżutka and Cook 2004). Additionally, the inactivation mechanisms of viruses is strictly related to the viral structure, where double-stranded DNA or RNA viruses, such as *Human adenovirus* and *Rotavirus*, have shown to be more resistant to UV radiation than viruses with single-stranded genetic material (Calgua et al. 2014). This characteristic is due to the capability to repair their genomes during replication.

On the other hand, enveloped viruses are more vulnerable to inactivation in the environment, due to the presence of solvents, detergents and disinfectants in wastewater that they destroy the viral envelope (Corpuz et al. 2020). However, it is worth mentioning that it is not a rule that all the viruses that present fecal-oral routes are non-enveloped. The presence of enveloped viruses, such as the coronavirus group have been reported in feces, once they present the replication cycle on the enteric tract (Christoff et al. 2021, Leung et al. 2003). In this sense, the importance of wastewater surveillance for these viruses have also been reported since a long time ago (Gundy et al. 2009).

Although all these microorganisms are of interest, especially considering the One Health concept, it is not possible to monitor all of them. Furthermore, researchers are always looking for indicators to be used for tracking wastewater contamination in the environment (Symonds and Breitbart 2015). Along the history of surveillance of microbial contamination, the most used indicators were the fecal indicator bacteria (FIB) such as *Escherichia coli, Enterococcus* sp. However, over time it was observed that there was no correlation between the presence of these fecal indicator bacteria and the enteric virus in the environmental sample matrices such as water, sediments and food (Carducci et al. 2009), due to the fact that bacteria are less resistant to wastewater treatments and less persistent in the environment.

Considering this, researchers have looked for virus indicator candidates, index and model organisms. The ideal characteristics for a viral indicator of wastewater-contamination is that it should be always present in contaminated environments, easy to detect and/or cultivate, and as resistant as the target pathogens (Farkas et al. 2020). Additionally, it is recommended to use contamination markers that are able to distinguish between animal- and human-derived pollution (Scott et al. 2002).

Enteric viruses are promising Microbial Source Tracking (MST), and in this sense, the *Human adenovirus* (HAdV) has been suggested as human fecal indicator (Hundesa et al. 2006), once they show high resistance to environmental conditions such as those related to pH, temperature and radiations (Carter 2005). Besides, double-stranded DNA such as HAdV present prolonged persistence on environmental samples if compared to RNA viruses (Mena and Gerba 2008), *Porcine circovirus* type 2 (PCV2) has been suggested as swine manure contamination for the same resistance profile (Viancelli et al. 2013), while *Bovine enterovirus* (BEV) as bovine feces indicator (Ley et al. 2002).

In this line of thought, researchers have investigated plant viruses as indicators of human fecal contamination. One example is the *Pepper Mild Mottle virus* (PMMoV), a non-enveloped, RNA virus, which is the most abundant virus that infects pepper plants (*Capsicum* spp.). Due to the wide use of pepper *in natura* and as a spice in processed food products, this virus is excreted on large amounts in feces of healthy human populations, in concentration of up to 10^9 virions per gram of dry weight fecal matter (Colson et al. 2010), but rarely found in animal feces (Kitajima et al. 2018). Additionally, they also present a geographically widespread distribution (Zhang et al. 2005).

Other groups of viruses that have been studied and suggested as human contamination indicators are the bacteriophages. About the discovery of bacteriophages, Letarov (2020) wrote a bright and enjoyable to read review (that we really invite the readers to spend time on), where he describes the successful, but almost unknown, history of Frederick William Twort, who had described the existence of a virus able to infect bacteria. However, due to Twort's professional choices, his article, published in 1915 in the journal *Lancet,* remained lost and the studies of other scientists are remembered by the phage description, Félix d'Hérelle in 1917 (Letarov 2020). d'Hérelle observed an unexplained phenomenon on bacteria culture when he was studying biological methods to control locust infestation. The coining of the term "bacteriophage" is also attributed to him, i.e., the description of bacteria that acquired resistance to phage and the investigation of these viruses on environmental samples. From that time to now, it is estimated that there are about 10^{31} particles of phages species in the planet (Hendrix 2003), varying greatly in morphology and replicative characteristics, with 96% of the identified phages belonging to the order *Caudovirales*, presenting double stranded DNA and tail, classified in the families of *Siphoviridae, Myoviridae* and *Podoviridae* (Dion et al. 2020, Sharp 2001).

In the 20s', phage therapy was a promising idea to treat bacterial infections, but the discovery of antibiotics let it fall apart (Viertel et al. 2014). However, in the last decade the wide and fast spread of antibiotic resistant bacteria made scientists bring back the idea of applying phage therapy (Golkar et al. 2014). Phages can be species specific, and can be a universal and safe technology that could be used to control infections of a special group of bacteria (Sieiro et al. 2020).

Considering these, and the number of phages on Earth, the surveillance for new phages for therapeutic purposes have been investigated, especially in matrices rich in bacteria such as wastewater from humans and animals (Weber-Dąbrowska et al. 2016). The investigation in this field is resumed in collecting environmental samples, removing the unwanted microorganisms and then placing this sample on a permissive bacterial host to observe the characteristic plaques lyse formation (Gill and Hyman 2010). Although it seems like a fantastic alternative, it is important to be careful to select a phage that does not present adverse effects, such as lysogenic life cycle, in which the virus may integrate its DNA in the bacterial genome and incorporate resistance genes on it (Kakasis and Panitsa 2019). Another important point is the use of phage on gram-negative bacteria, is the fact that after the bacteria cell rupture' there is the liberation of endotoxins (Kakasis and Panitsa 2019).

As the studies advance and improve the knowledge about the behavior of bacteriophage in different situations, they also increase the commercial options containing bacteriophages (Sulakvelidze 2013, Vikram et al. 2020). The use of some phage products were approved by the United States Department of Agriculture (USDA) to be used as food sanitizers such as Salmonelex™, ListShield™, ListShield™ (Rogovski et al. 2021a). In the food engineering research field, the bacteriophage application has been studied for biofilm control on the food industry and in the food production chain as a substitute for antibiotic application on fish, poultry, shrimps, oysters, sheep and swine production (Jessen and Lammert 2003, Moye et al. 2018).

The use of phage has also been explored in environmental scenarios, such as the application of lytic phages to reduce the prevalence of pathogenic *Salmonella* strains on poultry litter used in poultry production environments (Rogovski et al. 2021b). Additionally, the phage named MS2 has been suggested as a good operational monitoring indicator for wastewater treatment plants (Amarasiri et al. 2017). Phages infecting *E. coli,* called somatic coliphages and F-specific phages, have been used as wastewater contamination biomarkers (Farkas et al. 2020). In recent years, a new group called crass-like phages was discovered as part of human normal gut virome, and also has been suggested as an environmental contamination biomarker (Edwards et al. 2019).

2. About the future

Before ending this introduction about the fantastic virus' world, it is worth mentioning some not so old discoveries that are changing the way we deal with host-virus interaction. The first one is about the breaking of the virus' definition concept occurred in the 90's. At that time Scola et al. (2003) discovered the first amoeba giant viruses named Mimivirus and Mamavirus. These viruses were about 700 nm and unable to be filtered. This was against the knowledge that all viruses were filterable. Breaking the ancient concept and definition of "virus" as a filterable particle.

The second important point about the virus distribution in environments around the world that needs special attention is the transport by insects, such as houseflies, beetles and bees (Cunningham et al. 2022, Young et al. 2021). Once these insects are exposed to different contaminants during foraging activities, they could transport them to different places (McArt et al. 2014).

The third point is the application of new tools to identify and classify the viruses. The Quantitative Microbial Risk Analysis (QRMA) is a modeling approach to simulate and evaluate the risk associated with the exposure of humans and other animals to these pathogens (Zhang et al. 2019). The use of this methodology is still in its infancy compared to all other methodologies, but with important contributions to environmental virology (Emilse et al. 2021, Pasalari et al. 2019). Another important technique is the Next Generation Sequencing, cited before, that is associated with bioinformatics analysis. With these kind of approaches, it is possible to understand all the microbiome from a place, such as reclaimed water (Rosario et al. 2009), rivers (Lu et al. 2022), human diarrhea (Finkbeiner et al. 2008), human gut (Aggarwala et al. 2017), body virome (Bai et al. 2022), and to trace the profile of important viruses and pathogens found in these environments, so that the biology of these microorganisms, together with the way in which they are transmitted, can be elucidated.

However, the environmental contamination by viruses is a consequence of human activities. Thus, while the high population density continues to increase, resulting in a rapid, inadequate and unplanned urbanization, associated with the lack of sanitation, the wastewater will continue to input pathogens on the environment. Deforestation and destruction of natural habitats could bring back emerging and reemerging viruses (Bastaraud et al. 2020).

With these highlight of virology history, we would like to point out and motivate researchers about the abundant knowledge that can be gained about a large group of organisms on Earth. A huge amount of discoveries are still waiting to be made.

References

Adefisoye, M.A., Nwodo, U.U., Green, E. and Okoh, A.I. 2016. Quantitative PCR detection and characterization of Human Adenovirus, Rotavirus and Hepatitis A Virus in discharged effluents of two wastewater treatment facilities in the Eastern Cape, South Africa. Food Environ. Virol. 8: 262–274. https://doi.org/10.1007/s12560-016-9246-4.

Aggarwala, V., Liang, G. and Bushman, F.D. 2017. Viral communities of the human gut: metagenomic analysis of composition and dynamics. Mob. DNA 8: 12. https://doi.org/10.1186/s13100-017-0095-y.

Ahmed, A.T., El Gohary, F., Tzanakakis, V.A. and Angelakis, A.N. 2020. Egyptian and Greek water cultures and hydro-technologies in Ancient Times. Sustainability 12: 9760. https://doi.org/10.3390/su12229760.

Amarasiri, M., Kitajima, M., Nguyen, T.H., Okabe, S. and Sano, D. 2017. Bacteriophage removal efficiency as a validation and operational monitoring tool for virus reduction in wastewater reclamation: Review. Water Res. 121: 258–269. https://doi.org/10.1016/j.watres.2017.05.035.

Aw, T.G. and Rose, J.B. 2012. Detection of pathogens in water: from phylochips to qPCR to pyrosequencing. Curr. Opin. Biotechnol. 23: 422–430. https://doi.org/10.1016/j.copbio.2011.11.016.

Bai, G.-H., Lin, S.-C., Hsu, Y.-H. and Chen, S.-Y. 2022. The human virome: viral metagenomics, relations with human diseases, and therapeutic applications. Viruses 14: 278. https://doi.org/10.3390/v14020278.

Balali, G.I., Yar, D.D., Afua Dela, V.G. and Adjei-Kusi, P. 2020. Microbial contamination, an increasing threat to the consumption of fresh fruits and vegetables in Today's World. Int. J. Microbiol. 2020: 1–13. https://doi.org/10.1155/2020/3029295.

Bastaraud, A., Cecchi, P., Handschumacher, P., Altmann, M. and Jambou, R. 2020. Urbanization and waterborne pathogen emergence in low-income countries: where and how to conduct surveys? Int. J. Environ. Res. Public Health 17: 480. https://doi.org/10.3390/ijerph17020480.

Bellou, M., Kokkinos, P. and Vantarakis, A. 2013. Shellfish-borne viral outbreaks: a systematic review. Food Environ. Virol. 5: 13–23. https://doi.org/10.1007/s12560-012-9097-6.

Bibby, K., Viau, E. and Peccia, J. 2011. Viral metagenome analysis to guide human pathogen monitoring in environmental samples. Lett. Appl. Microbiol. 52: 386–392. https://doi.org/10.1111/j.1472-765X.2011.03014.x.

Bosch, A., Gkogka, E., Le Guyader, F.S., Loisy-Hamon, F., Lee, A., van Lieshout, L. et al. 2018. Foodborne viruses: Detection, risk assessment, and control options in food processing. Int. J. Food Microbiol. 285: 110–128. https://doi.org/10.1016/j.ijfoodmicro.2018.06.001.

Bozkurt, H., Phan-Thien, K.-Y., van Ogtrop, F., Bell, T. and McConchie, R. 2021. Outbreaks, occurrence, and control of norovirus and hepatitis a virus contamination in berries: A review. Crit. Rev. Food Sci. Nutr. 61: 116–138. https://doi.org/10.1080/10408398.2020.1719383.

Butt, A.A., Aldridge, K.E. and Sanders, C.V. 2004. Infections related to the ingestion of seafood Part I: viral and bacterial infections. Lancet Infect. Dis. 4: 201–212. https://doi.org/10.1016/S1473-3099(04)00969-7.

Cadamuro, R.D., Viancelli, A., Michelon, W., Fonseca, T.G., Mass, A.P., Krohn, D.M.A. et al. 2021. Enteric viruses in lentic and lotic freshwater habitats from Brazil's Midwest and South regions in the Guarani Aquifer area. Environ. Sci. Pollut. Res. 28: 31653–31658. https://doi.org/10.1007/s11356-021-13029-y.

Calgua, B., Carratalà, A., Guerrero-Latorre, L., de Abreu Corrêa, A., Kohn, T., Sommer, R. et al. 2014. UVC inactivation of dsDNA and ssRNA viruses in water: UV Fluences and a qPCR-based approach to evaluate decay on viral infectivity. Food Environ. Virol. 6: 260–268. https://doi.org/10.1007/s12560-014-9157-1.

Carducci, A., Verani, M., Battistini, R., Pizzi, F., Rovini, E., Andreoli, E. et al. 2006. Epidemiological surveillance of human enteric viruses by monitoring of different environmental matrices. Water Sci. Technol. 54: 239–244. https://doi.org/10.2166/wst.2006.475.

Carducci, A., Battistini, R., Rovini, E. and Verani, M. 2009. Viral removal by wastewater treatment: monitoring of indicators and pathogens. Food Environ. Virol. 1: 85–91. https://doi.org/10.1007/s12560-009-9013-x.

Carter, M.J. 2005. Enterically infecting viruses: pathogenicity, transmission and significance for food and waterborne infection. J. Appl. Microbiol. 98: 1354–1380. https://doi.org/10.1111/j.1365-2672.2005.02635.x.

Christoff, A.P., Cruz, G.N.F., Sereia, A.F.R., Boberg, D.R., de Bastiani, D.C., Yamanaka, L.E. et al. 2021. Swab pooling: A new method for large-scale RT-qPCR screening of SARS-CoV-2 avoiding sample dilution. PLoS One 16: e0246544. https://doi.org/10.1371/journal.pone.0246544.

Chuttani, H.K., Sidhu, A.S., Wig, K.L., Gupta, D.N. and Ramalingaswami, V. 1966. Follow-up study of cases from the delhi epidemic of infectious hepatitis of 1955–6. BMJ 2: 676–679. https://doi.org/10.1136/bmj.2.5515.676.

Colson, P., Richet, H., Desnues, C., Balique, F., Moal, V., Grob, J.-J. et al. 2010. Pepper mild mottle virus, a plant virus associated with specific immune responses, fever, abdominal pains, and pruritus in humans. PLoS One 5: e10041. https://doi.org/10.1371/journal.pone.0010041.

Corpuz, M.V.A., Buonerba, A., Vigliotta, G., Zarra, T., Ballesteros, F., Campiglia, P. et al. 2020. Viruses in wastewater: occurrence, abundance and detection methods. Sci. Total Environ. 745: 140910. https://doi.org/10.1016/j.scitotenv.2020.140910.

Cunningham, M.M., Tran, L., McKee, C.G., Ortega Polo, R., Newman, T., Lansing, L. et al. 2022. Honey bees as biomonitors of environmental contaminants, pathogens, and climate change. Ecol. Indic. 134: 108457. https://doi.org/10.1016/j.ecolind.2021.108457.

de Matos Nascimento, A., de Paula, V.R., Dias, E.H.O., da Costa Carneiro, J. and Otenio, M.H. 2020. Quantitative microbial risk assessment of occupational and public risks associated with bioaerosols generated during the application of dairy cattle wastewater as biofertilizer. Sci. Total Environ. 745: 140711. https://doi.org/10.1016/j.scitotenv.2020.140711.

Destoumieux-Garzón, D., Mavingui, P., Boetsch, G., Boissier, J., Darriet, F., Duboz, P. et al. 2018. The One Health Concept: 10 Years old and a long road ahead. Front. Vet. Sci. 5. https://doi.org/10.3389/fvets.2018.00014.

DiCaprio, E., Purgianto, A. and Li, J. 2015. Effects of abiotic and biotic stresses on the internalization and dissemination of human norovirus surrogates in growing romaine lettuce. Appl. Environ. Microbiol. 81: 4791–4800. https://doi.org/10.1128/AEM.00650-15.

Dion, M.B., Oechslin, F. and Moineau, S. 2020. Phage diversity, genomics and phylogeny. Nat. Rev. Microbiol. 18: 125–138. https://doi.org/10.1038/s41579-019-0311-5.

Edwards, R.A., Vega, A.A., Norman, H.M., Ohaeri, M., Levi, K., Dinsdale, E.A. et al. 2019. Global phylogeography and ancient evolution of the widespread human gut virus crAssphage. Nat. Microbiol. 4: 1727–1736. https://doi.org/10.1038/s41564-019-0494-6.

Emilse, P.V., Matías, V., Cecilia, M.L., Oscar, G.M., Gisela, M., Guadalupe, D. et al. 2021. Enteric virus presence in green vegetables and associated irrigation waters in a rural area from Argentina. A quantitative microbial risk assessment. LWT 144: 111201. https://doi.org/10.1016/j.lwt.2021.111201.

Farkas, K., Walker, D.I., Adriaenssens, E.M., McDonald, J.E., Hillary, L.S., Malham, S.K. et al. 2020. Viral indicators for tracking domestic wastewater contamination in the aquatic environment. Water Res. 181: 115926. https://doi.org/10.1016/j.watres.2020.115926.

Federigi, I., Verani, M., Donzelli, G., Cioni, L. and Carducci, A. 2019. The application of quantitative microbial risk assessment to natural recreational waters: A review. Mar. Pollut. Bull. 144: 334–350. https://doi.org/10.1016/j.marpolbul.2019.04.073.

Finkbeiner, S.R., Allred, A.F., Tarr, P.I., Klein, E.J., Kirkwood, C.D. and Wang, D. 2008. Metagenomic analysis of human diarrhea: viral detection and discovery. PLoS Pathog. 4: e1000011. https://doi.org/10.1371/journal.ppat.1000011.

Fong, T.-T. and Lipp, E.K. 2005. Enteric viruses of humans and animals in aquatic environments: health risks, detection, and potential water quality assessment tools. Microbiol. Mol. Biol. Rev. 69: 357–371. https://doi.org/10.1128/MMBR.69.2.357-371.2005.

Fongaro, G., Padilha, J., Schissi, C.D., Nascimento, M.A., Bampi, G.B., Viancelli, A. et al. 2015. Human and animal enteric virus in groundwater from deep wells, and recreational and network water. Environ. Sci. Pollut. Res. 22: 20060–20066. https://doi.org/10.1007/s11356-015-5196-x.

Fornari, B.F., Nicodem, L.F., Fonseca, T.G., Rossi, P., Knoblauch, P.M., Reis, P. et al. 2021. Water contamination by enteric virus and superbugs in rural areas and the implications in the One Health context. Int. J. Environ. Stud. 78: 785–796. https://doi.org/10.1080/00207233.2020.1842001.

Fuzawa, M., Duan, J., Shisler, J.L. and Nguyen, T.H. 2021. Peracetic acid sanitation on arugula microgreens contaminated with surface-attached and internalized Tulane Virus and Rotavirus. Food Environ. Virol. 13: 401–411. https://doi.org/10.1007/s12560-021-09473-1.

Galbraith, D.W., Anderson, M.T. and Herzenberg, L.A. 1998. Chapter 19: Flow Cytometric Analysis and FACS Sorting of Cells Based on GFP Accumulation. pp. 315–341. https://doi.org/10.1016/S0091-679X(08)61963-9.

Ganguli, P.S., Chen, W. and Yates, M.V. 2011. Detection of murine norovirus-1 by using TAT peptide-delivered molecular beacons. Appl. Environ. Microbiol. 77: 5517–5520. https://doi.org/10.1128/AEM.03048-10.

Gerba, C.P., Betancourt, W.Q. and Kitajima, M. 2017. How much reduction of virus is needed for recycled water: A continuous changing need for assessment? Water Res. 108: 25–31. https://doi.org/10.1016/j.watres.2016.11.020.

Gibbs, E.P.J. 2014. The evolution of One Health: a decade of progress and challenges for the future. Vet. Rec. 174: 85–91. https://doi.org/10.1136/vr.g143.

Gill, J. and Hyman, P. 2010. Phage choice, isolation, and preparation for phage therapy. Curr. Pharm. Biotechnol. 11: 2–14. https://doi.org/10.2174/138920110790725311.

Girones, R. and Bofill-Mas, S. 2013. Virus indicators for food and water. pp. 483–509. *In*: Viruses in Food and Water. Elsevier. https://doi.org/10.1533/9780857098870.4.483.

Golkar, Z., Bagasra, O. and Pace, D.G. 2014. Bacteriophage therapy: a potential solution for the antibiotic resistance crisis. J. Infect. Dev. Ctries. 8: 129–136. https://doi.org/10.3855/jidc.3573.

Gundy, P.M., Gerba, C.P. and Pepper, I.L. 2009. Survival of coronaviruses in water and wastewater. Food Environ. Virol. 1: 10. https://doi.org/10.1007/s12560-008-9001-6.

Hall, E.L. and Dietrich, A.M. 2000. A brief history of drinking water. Opflow 26: 46–49. https://doi. org/10.1002/j.1551-8701.2000.tb02243.x.

Hejkal, T.W., Wellings, F.M., Lewis, A.L. and LaRock, P.A. 1981. Distribution of viruses associated with particles in waste water. Appl. Environ. Microbiol. 41: 628–634. https://doi.org/10.1128/ aem.41.3.628-634.1981.

Hendrix, R.W. 2003. Bacteriophage genomics. Curr. Opin. Microbiol. 6: 506–511. https://doi. org/10.1016/j.mib.2003.09.004.

Hernandez-Morga, J., Leon-Felix, J., Peraza-Garay, F., Gil-Salas, B.G. and Chaidez, C. 2009. Detection and characterization of hepatitis A virus and Norovirus in estuarine water samples using ultrafiltration— RT-PCR integrated methods. J. Appl. Microbiol. 106: 1579–1590. https://doi.org/10.1111/j.1365-2672.2008.04125.x.

Hess, T. and Sutcliffe, C. 2018. The exposure of a fresh fruit and vegetable supply chain to global water-related risks. Water Int. 43: 746–761. https://doi.org/10.1080/02508060.2018.1515569.

Hirneisen, K.A., Sharma, M. and Kniel, K.E. 2012. Human enteric pathogen internalization by root uptake into food crops. Foodborne Pathog. Dis. 9: 396–405. https://doi.org/10.1089/fpd.2011.1044.

Hlavsa, M.C., Aluko, S.K., Miller, A.D., Person, J., Gerdes, M.E., Lee, S. et al. 2021. Outbreaks associated with treated recreational water—United States, 2015–2019. MMWR. Morb. Mortal. Wkly. Rep. 70: 733–738. https://doi.org/10.15585/mmwr.mm7020a1.

Hofstra, N., Bouwman, A.F., Beusen, A.H.W. and Medema, G.J. 2013. Exploring global Cryptosporidium emissions to surface water. Sci. Total Environ. 442: 10–19. https://doi.org/10.1016/j. scitotenv.2012.10.013.

Hofstra, N., Vermeulen, L.C., Derx, J., Flörke, M., Mateo-Sagasta, J., Rose, J. et al. 2019. Priorities for developing a modelling and scenario analysis framework for waterborne pathogen concentrations in rivers worldwide and consequent burden of disease. Curr. Opin. Environ. Sustain. 36: 28–38. https://doi.org/10.1016/j.cosust.2018.10.002.

Hundesa, A., Maluquer de Motes, C., Bofill-Mas, S., Albinana-Gimenez, N. and Girones, R. 2006. Identification of human and animal adenoviruses and polyomaviruses for determination of sources of fecal contamination in the environment. Appl. Environ. Microbiol. 72: 7886–7893. https://doi. org/10.1128/AEM.01090-06.

Ikner, L.A., Gerba, C.P. and Bright, K.R. 2012. Concentration and recovery of viruses from water: a comprehensive review. Food Environ. Virol. 4: 41–67. https://doi.org/10.1007/s12560-012-9080-2.

Jessen, B. and Lammert, L. 2003. Biofilm and disinfection in meat processing plants. Int. Biodeterior. Biodegradation 51: 265–269. https://doi.org/10.1016/S0964-8305(03)00046-5.

Kakasis, A. and Panitsa, G. 2019. Bacteriophage therapy as an alternative treatment for human infections. A comprehensive review. Int. J. Antimicrob. Agents 53: 16–21. https://doi.org/10.1016/j. ijantimicag.2018.09.004.

Keswick, B.H., Gerba, C.P., DuPont, H.L. and Rose, J.B. 1984. Detection of enteric viruses in treated drinking water. Appl. Environ. Microbiol. 47: 1290–1294. https://doi.org/10.1128/aem.47.6.1290-1294.1984.

Kitajima, M., Sassi, H.P. and Torrey, J.R. 2018. Pepper mild mottle virus as a water quality indicator. npj Clean Water 1: 19. https://doi.org/10.1038/s41545-018-0019-5.

Kokkinos, P., Kozyra, I., Lazic, S., Söderberg, K., Vasickova, P., Bouwknegt, M. et al. 2017. Virological quality of irrigation water in leafy green vegetables and berry fruits production chains. Food Environ. Virol. 9: 72–78. https://doi.org/10.1007/s12560-016-9264-2.

Kolawole, A.O. and Wobus, C.E. 2020. Gastrointestinal organoid technology advances studies of enteric virus biology. PLOS Pathog. 16: e1008212. https://doi.org/10.1371/journal.ppat.1008212.

Koopmans, M. and Duizer, E. 2004. Foodborne viruses: an emerging problem. Int. J. Food Microbiol. 90: 23–41. https://doi.org/10.1016/S0168-1605(03)00169-7.

Kurdziel, A.S., Wilkinson, N., Langton, S. and Cook, N. 2001. Survival of poliovirus on soft fruit and salad vegetables. J. Food Prot. 64: 706–709. https://doi.org/10.4315/0362-028X-64.5.706.

Lecoq, H. 2001. Découverte du premier virus, le virus de la mosaïque du tabac : 1892 ou 1898 ? Comptes Rendus l'Académie des Sci. - Ser. III - Sci. la Vie 324: 929–933. https://doi.org/10.1016/S0764-4469(01)01368-3.

Letarov, A.V. 2020. History of early bacteriophage research and emergence of key concepts in virology. Biochem. 85: 1093–1112. https://doi.org/10.1134/S0006297920090096.

Leung, W.K., To, K., Chan, P.K., Chan, H.L., Wu, A.K., Lee, N. et al. 2003. Enteric involvement of severe acute respiratory syndrome-associated coronavirus infection. Gastroenterology 125: 1011–1017. https://doi.org/10.1016/j.gastro.2003.08.001.

Lévesque, A., Jubinville, E., Hamon, F. and Jean, J. 2021. Detection of enteric viruses on strawberries and raspberries using capture by Apolipoprotein H. Foods 10: 3139. https://doi.org/10.3390/foods10123139.

Ley, V., Higgins, J. and Fayer, R. 2002. Bovine enteroviruses as indicators of fecal contamination. Appl. Environ. Microbiol. 68: 3455–3461. https://doi.org/10.1128/AEM.68.7.3455-3461.2002.

Lu, J., Yang, S., Zhang, X., Tang, X., Zhang, J., Wang, X. et al. 2022. Metagenomic analysis of viral community in the Yangtze River expands known eukaryotic and prokaryotic virus diversity in freshwater. Virol. Sin. 37: 60–69. https://doi.org/10.1016/j.virs.2022.01.003.

Masclaux, F.G., Hotz, P., Friedli, D., Savova-Bianchi, D. and Oppliger, A. 2013. High occurrence of hepatitis E virus in samples from wastewater treatment plants in Switzerland and comparison with other enteric viruses. Water Res. 47: 5101–5109. https://doi.org/10.1016/j.watres.2013.05.050.

McArt, S.H., Koch, H., Irwin, R.E. and Adler, L.S. 2014. Arranging the bouquet of disease: floral traits and the transmission of plant and animal pathogens. Ecol. Lett. 17: 624–636. https://doi.org/10.1111/ele.12257.

Mcleod, C., Hay, B., Grant, C., Greening, G. and Day, D. 2009. Localization of norovirus and poliovirus in Pacific oysters. J. Appl. Microbiol. 106: 1220–1230. https://doi.org/10.1111/j.1365-2672.2008.04091.x.

Melnick, J.L. 1947. Poliomyelitis virus in urban sewage in epidemic and in nonepidemic times. Am. J. Epidemiol. 45: 240–253. https://doi.org/10.1093/oxfordjournals.aje.a119132.

Mena, K.D. and Gerba, C.P. 2008. Waterborne adenovirus. *In*: Reviews of Environmental Contamination and Toxicology Volume 198. Springer New York, New York, NY, pp. 1–35. https://doi.org/10.1007/978-0-387-09647-6_4.

Metcalf, T.G., Melnick, J.L. and Estes, M.K. 1995. Environmental virology: from detection of virus in sewage and water by isolation to identification by molecular biology—a trip of over 50 years. Annu. Rev. Microbiol. 49: 461–487. https://doi.org/10.1146/annurev.mi.49.100195.002333.

Miagostovich, M.P. and Vieira, C.B. 2017. Environmental virology. pp. 81–117. *In*: Human Virology in Latin America. Springer International Publishing, Cham, https://doi.org/10.1007/978-3-319-54567-7_6.

Moresco, V., Viancelli, A., Nascimento, M.A., Souza, D.S.M., Ramos, A.P.D. et al. 2012. Microbiological and physicochemical analysis of the coastal waters of southern Brazil. Mar. Pollut. Bull. 64: 40–48. https://doi.org/10.1016/j.marpolbul.2011.10.026.

Moye, Z., Woolston, J. and Sulakvelidze, A. 2018. Bacteriophage applications for food production and processing. Viruses 10: 205. https://doi.org/10.3390/v10040205.

Pasalari, H., Ataei-Pirkooh, A., Aminikhah, M., Jafari, A.J. and Farzadkia, M. 2019. Assessment of airborne enteric viruses emitted from wastewater treatment plant: Atmospheric dispersion model, quantitative microbial risk assessment, disease burden. Environ. Pollut. 253: 464–473. https://doi.org/10.1016/j.envpol.2019.07.010.

Pepper, I.L., Brooks, J.P. and Gerba, C.P. 2019. Land application of organic residuals: municipal biosolids and animal manures. pp. 419–434. *In*: Environmental and Pollution Science. Elsevier. https://doi.org/10.1016/B978-0-12-814719-1.00023-9.

Polo, D., Varela, M.F. and Romalde, J.L. 2015. Detection and quantification of hepatitis A virus and norovirus in Spanish authorized shellfish harvesting areas. Int. J. Food Microbiol. 193: 43–50. https://doi.org/10.1016/j.ijfoodmicro.2014.10.007.

Raina, S.A., Bhat, R.A., Qadri, H. and Dutta, A. 2020. Values of biofertilizers for sustainable management in agricultural industries. pp. 121–137. *In*: Bioremediation and Biotechnology, Vol 2. Springer International Publishing, Cham. https://doi.org/10.1007/978-3-030-40333-1_7.

Richert-Pöggeler, K.R., Franzke, K., Hipp, K. and Kleespies, R.G. 2019. Electron microscopy methods for virus diagnosis and high resolution analysis of viruses. Front. Microbiol. 9. https://doi.org/10.3389/fmicb.2018.03255.

Rodríguez-Lázaro, D., Cook, N., Ruggeri, F.M., Sellwood, J., Nasser, A., Nascimento, M.S.J. et al. 2012. Virus hazards from food, water and other contaminated environments. FEMS Microbiol. Rev. 36: 786–814. https://doi.org/10.1111/j.1574-6976.2011.00306.x.

Rogovski, P., Cadamuro, R.D., da Silva, R., de Souza, E.B., Bonatto, C., Viancelli, A. et al. 2021a. Uses of bacteriophages as bacterial control tools and environmental safety indicators. Front. Microbiol. 12. https://doi.org/10.3389/fmicb.2021.793135.

Rogovski, P., Silva, R. da, Cadamuro, R.D., Souza, E.B. de, Savi, B.P., Viancelli, A. et al. 2021b. Salmonella enterica Serovar Enteritidis control in poultry litter mediated by lytic bacteriophage isolated from swine manure. Int. J. Environ. Res. Public Health 18: 8862. https://doi.org/10.3390/ijerph18168862.

Roos, Y.H. 2020. Water and pathogenic viruses inactivation—food engineering perspectives. Food Eng. Rev. 12: 251–267. https://doi.org/10.1007/s12393-020-09234-z.

Rosario, K., Nilsson, C., Lim, Y.W., Ruan, Y. and Breitbart, M. 2009. Metagenomic analysis of viruses in reclaimed water. Environ. Microbiol. 11: 2806–2820. https://doi.org/10.1111/j.1462-2920.2009.01964.x.

Rzeżutka, A. and Cook, N. 2004. Survival of human enteric viruses in the environment and food. FEMS Microbiol. Rev. 28: 441–453. https://doi.org/10.1016/j.femsre.2004.02.001.

Sánchez, G. and Bosch, A. 2016. Survival of enteric viruses in the environment and food. pp. 367–392. *In*: Viruses in Foods. Springer International Publishing, Cham. https://doi.org/10.1007/978-3-319-30723-7_13.

Sassi, H.P., van Ogtrop, F., Morrison, C.M., Zhou, K., Duan, J.G. and Gerba, C.P. 2020. Sediment re-suspension as a potential mechanism for viral and bacterial contaminants. J. Environ. Sci. Heal. Part A 55: 1398–1405. https://doi.org/10.1080/10934529.2020.1796118.

Savini, F., Giacometti, F., Tomasello, F., Pollesel, M., Piva, S., Serraino, A. and De Cesare, A. 2021. Assessment of the impact on human health of the presence of norovirus in bivalve molluscs: what data do we miss? Foods 10: 2444. https://doi.org/10.3390/foods10102444.

Scholthof, K.-B.G. 2004. Tobacco mosaic virus: A model system for plant biology. Annu. Rev. Phytopathol. 42: 13–34. https://doi.org/10.1146/annurev.phyto.42.040803.140322.

Scola, B. La, Audic, S., Robert, C., Jungang, L., de Lamballerie, X., Drancourt, M. et al. 2003. A giant virus in Amoebae. Science (80-.). 299: 2033–2033. https://doi.org/10.1126/science.1081867.

Scott, T.M., Rose, J.B., Jenkins, T.M., Farrah, S.R. and Lukasik, J. 2002. Microbial source tracking: current methodology and future directions. Appl. Environ. Microbiol. 68: 5796–5803. https://doi.org/10.1128/AEM.68.12.5796-5803.2002.

Seitz, S.R., Leon, J.S., Schwab, K.J., Lyon, G.M., Dowd, M., McDaniels, M. et al. 2011. Norovirus infectivity in humans and persistence in water. Appl. Environ. Microbiol. 77: 6884–6888. https://doi.org/10.1128/AEM.05806-11.

Sharp, R. 2001. Bacteriophages: biology and history. J. Chem. Technol. Biotechnol. 76: 667–672. https://doi.org/10.1002/jctb.434.

Sieiro, C., Areal-Hermida, L., Pichardo-Gallardo, Á., Almuiña-González, R., de Miguel, T., Sánchez, S. et al. 2020. A hundred years of bacteriophages: can phages replace antibiotics in agriculture and aquaculture? Antibiotics 9: 493. https://doi.org/10.3390/antibiotics9080493.

Sobral Marques Souza, D., Miura, T., Le Mennec, C., Barardi, C.R.M. and Le Guyader, F.S. 2015. Retention of rotavirus infectivity in mussels heated by using the french recipe Moules Marinières. J. Food Prot. 78: 2064–2069. https://doi.org/10.4315/0362-028X.JFP-15-191.

Souza, D.S.M., Dominot, A.F.Á., Moresco, V. and Barardi, C.R.M. 2018. Presence of enteric viruses, bioaccumulation and stability in Anomalocardia brasiliana clams (Gmelin, 1791). Int. J. Food Microbiol. 266: 363–371. https://doi.org/10.1016/j.ijfoodmicro.2017.08.004.

Stals, A., Baert, L., De Keuckelaere, A., Van Coillie, E. and Uyttendaele, M. 2011. Evaluation of a norovirus detection methodology for ready-to-eat foods. Int. J. Food Microbiol. 145: 420–425. https://doi.org/10.1016/j.ijfoodmicro.2011.01.013.

Sukla, S., Mondal, P., Biswas, S. and Ghosh, S. 2021. A rapid and easy-to-perform method of nucleic-acid based dengue virus diagnosis using fluorescence-based molecular beacons. Biosensors 11: 479. https://doi.org/10.3390/bios11120479.

Sulakvelidze, A. 2013. Using lytic bacteriophages to eliminate or significantly reduce contamination of food by foodborne bacterial pathogens. J. Sci. Food Agric. 93: 3137–3146. https://doi.org/10.1002/jsfa.6222.

Symonds, E.M. and Breitbart, M. 2015. Affordable enteric virus detection techniques are needed to support changing paradigms in water quality management. CLEAN - Soil, Air, Water 43: 8–12. https://doi.org/10.1002/clen.201400235.

United Nations. 2020. Ensure availability and sustainable management of water and sanitation for all. Available https//sdgs.un.org/goals/goal6. Accessed 18 July 2021.

Upfold, N.S., Luke, G.A. and Knox, C. 2021. Occurrence of human enteric viruses in water sources and shellfish: a focus on Africa. Food Environ. Virol. 13: 1–31. https://doi.org/10.1007/s12560-020-09456-8.

Viancelli, A., Kunz, A., Steinmetz, R.L.R., Kich, J.D., Souza, C.K., Canal, C.W. et al. 2013. Performance of two swine manure treatment systems on chemical composition and on the reduction of pathogens. Chemosphere 90: 1539–1544. https://doi.org/10.1016/j.chemosphere.2012.08.055.

Viertel, T.M., Ritter, K. and Horz, H.-P. 2014. Viruses versus bacteria—novel approaches to phage therapy as a tool against multidrug-resistant pathogens. J. Antimicrob. Chemother. 69: 2326–2336. https://doi.org/10.1093/jac/dku173.

Vikram, A., Woolston, J. and Sulakvelidze, A. 2020. Phage biocontrol applications in food production and processing. *In*: Bacterial Viruses: Exploitation for Biocontrol and Therapeutics. Caister Academic Press. https://doi.org/10.21775/9781913652517.08.

Warriner, K. 2005. Pathogens in vegetables. pp. 3–43. *In*: Improving the Safety of Fresh Fruit and Vegetables. Elsevier. https://doi.org/10.1533/9781845690243.1.3.

Weber-Dąbrowska, B., Jończyk-Matysiak, E., Żaczek, M., Łobocka, M., Łusiak-Szelachowska, M. and Górski, A. 2016. Bacteriophage procurement for therapeutic purposes. Front. Microbiol. 7. https://doi.org/10.3389/fmicb.2016.01177.

Wei, J., Jin, Y., Sims, T. and Kniel, K.E. 2011. Internalization of Murine Norovirus 1 by Lactuca sativa during irrigation. Appl. Environ. Microbiol. 77: 2508–2512. https://doi.org/10.1128/AEM.02701-10.

Wheeler, C., Vogt, T.M., Armstrong, G.L., Vaughan, G., Weltman, A., Nainan, O.V. et al. 2005. An outbreak of Hepatitis A associated with green onions. N. Engl. J. Med. 353: 890–897. https://doi.org/10.1056/NEJMoa050855.

WHO. 2022. World Health Organization - The top 10 causes of death. URL https://www.who.int/news-room/fact-sheets/detail/the-top-10-causes-of-death.

Wieczorek, P., Budziszewska, M., Frąckowiak, P. and Obrępalska-Stęplowska, A. 2020. Development of a New Tomato Torrado virus-based vector tagged with GFP for monitoring virus movement in plants. Viruses 12: 1195. https://doi.org/10.3390/v12101195.

Young, K.I., Valdez, F., Vaquera, C., Campos, C., Zhou, L., Vessels, H.K. et al. 2021. Surveillance along the Rio Grande during the 2020 Vesicular Stomatitis outbreak reveals spatio-temporal dynamics of and viral RNA detection in black flies. Pathogens 10: 1264. https://doi.org/10.3390/pathogens10101264.

Zhang, Q., Gallard, J., Wu, B., Harwood, V.J., Sadowsky, M.J., Hamilton, K.A. et al. 2019. Synergy between quantitative microbial source tracking (qMST) and quantitative microbial risk assessment (QMRA): A review and prospectus. Environ. Int. 130: 104703. https://doi.org/10.1016/j.envint.2019.03.051.

Zhang, T., Breitbart, M., Lee, W.H., Run, J.-Q., Wei, C.L., Soh, S.W.L. et al. 2005. RNA viral community in human feces: prevalence of plant pathogenic viruses. PLoS Biol. 4: e3. https://doi.org/10.1371/journal.pbio.0040003.

Zinsstag, J., Schelling, E., Waltner-Toews, D. and Tanner, M. 2015. One Health: The Theory and Practice of Integrated Health Approaches. CABI, Wallingford. https://doi.org/10.1079/9781780643410.0000.

Chapter 2
Enteroviruses
Impacts and Challengers

Beatriz Pereira Savi, Rafael Dorighello Cadamuro, Mariana Elois,
Giulia Von Tönnemann Pilati, Mariane Dahmer,
Helena Yurevna Caio, Júlia Zanette Penso,
Doris Sobral Marques Souza and *Gislaine Fongaro**

llllllllllllllllllllllllllllllllllllllllllllllllllllllllllllllllllllllllllllllllllllllllllllllllllllllllllllllllllllllllllllllllllllll

1. Introduction

1.1 History of Enteroviruses

The species of genus *Enterovirus* addressed in this chapter belong to the family *Picornaviridae*. The members of this taxon possess RNA as genome and their viral particles in general are small, which explains the name of the family (pico meaning small and RNA being their genome). Picornavirus virions present an icosahedral structure, non-enveloped, varying in size between 22–30 nm and RNA genome varying from 7.5 to 9 kb. The 5 subfamilies within this family are, *Caphtovirinae, Ensavirinae, Heptrevirinae, Kodimesavirinae,* and *Paavirinae*. The classification of genera is composed of comparison of morphology, physicochemical, biologic properties, antigenic structures, genomic sequence and mode of replication. The capsid is composed of 180 structural proteins, being 60 copies of VP1, VP2 and VP3. VP4 is located inside the viral capsid (Payne et al. 2017).

Poliovirus is an important viral particle representative among groups of Poliovirus, responsible for Poliomyelitis disease. Poliomyelitis was discovered and described by Michael Underwood in 1789. The first epidemic related to poliomyelitis in the United States was reported during the year 1843, after that some events occasionally happened until the 20th century when the improvement of public

Laboratory of Applied Virology, Federal University of Santa Catarina, Department of Microbiology, Immunology and Parasitology, Santa Catarina State, Brazil.
* Corresponding author: gislaine.fongaro@ufsc.br

sanitization altered the distribution, occurrence, age and chance of severity. During summers, epidemic outbreaks happen paralyzing hundreds of thousands of children and adults, across rich and poor countries (Tulchinsky 2018).

Poliovirus has 3 different strains in which the symptoms of infection can vary from asymptomatic to acute nonspecific mild febrile illness. Poliomyelitis is transmitted through oral-fecal contamination by direct person-to-person contact, but environmental contamination from sewage can also spread the virus (Tulchinsky 2018).

Poliovirus was grown in tissue culture for the first time in 1949, by John Enders, which allowed the development of several vaccines with inactivated viruses, such as the Inactivated Polio Vaccine (IPV) by Jonas Salk during 1950. In 1951, Salk determined that it has 3 different strains of Polio and developed a vaccine. The polio vaccine tests began in 1952, including a largest clinical trial involving 1.8 million children in the United States, Canada and Finland (Tulchinsky 2018).

In 1955, an incident led to inadequate production of inactivated viruses which were used to immunize 200,000 persons, of whom 70,000 became ill, 200 individuals were paralyzed and more than 10 died. After that, Salk vaccine was replaced with Sabin, which was an Oral Polio Vaccine (OPV). Sabin believed that an oral vaccine could interrupt the transmission and spread of polioviruses, and was approved by FDA and encouraged by the World Health Organization. Between 1955 and 1961 OPV was tested on 100 million people in Eastern Europe, Soviet Union, Singapore, Mexico and the Netherlands. OPV was licensed for tests in the US in 1960, recommended for general use by the American Medical Association in 1961 and adopted as the standard polio vaccine in 1963. The OPV possesses the benefits of effectiveness in the gut, low cost of production and non-invasiveness (Tulchinsky 2018).

Nonetheless, IPV and OPV have limitations. IPV provides a high production of humoral antibodies, but in return does not provide immunity to cells of the intestinal lining. Considering that wild *Poliovirus* (WPV) may not produce symptoms or clinical disease but can grow in the gut, causing the reintroduction of WPV in the environment. The production of IPV is costlier than OPV, at same time IPV requires a health professional to administer (Tulchinsky 2018).

OPV induces humoral and cellular immunity but rarely converts to wild status, as a result spreading in multiple outbreaks of clinical paralytic poliomyelitis. OPV can spread through sewage, considering the replication in the gut and excretion, which can help the process of immunization inside a community that possesses sanitary issues. OPV is safe and effective for immunization of children which helps eradicate polio (Tulchinsky 2018). It is composed of live attenuated polioviruses, but rarely genetic mutations can occur in the gut of an immunized person, generating the vaccine-associated paralytic polio (VAPP). In a few rare occasions, outbreaks of VAPP can spread as vaccine-derived poliovirus (VDPV) in approximately 1 in 2.7 million doses applied. Considering this risk, the US Advisory Committee on Immunization Practices (ACIP) recommended the revision of schedule to include two doses of IPV followed by two doses of OPV. Statistically the chance of emergence of VAPP is diminished with this schedule, being recommended application of IPV only as vaccinations strategy. OPV is a better choice when considering countries with less than 90 percent coverage in routine immunization due to its spread environmentally. The combination of OPV with IPV presents a great potential for completion of eradication (Tulchinsky 2018).

In 1988, World Health Organization (WHO) started the global polio eradication program called "Global Polio Eradication Initiative (GPEI)". In 2016, 85% of infants in the world received three doses of polio vaccine, whereas in 1990 only 70% received all doses. This could be observed for example in the African region, where the coverage with three doses (OPV) was 55% in 1990 and 73% in 2016. In India, the coverage of three doses (OPV) increased from 66% in 1990 to 86% in 2015 and the last reported polio case was in 2011. Since 1988, the number of polio cases has reduced by 99%, and, in 2016 only two countries, Afghanistan and Pakistan have not interrupted WPV, remaining endemic (Tulchinsky 2018). There are examples that raise concern about countries declared polio free experiencing the reentry of WPV and VAPP. India and Nigeria had no polio cases during 2011 and 2014, while Afghanistan and Pakistan showed a decrease in the number of WPV cases. In 2016, three different laboratories confirmed cases of WPV and one laboratory identified the circulation of VDVP (Tulchinsky 2018).

In most countries, a mechanism has been instituted to build effective surveillance and immunization systems and programs. This in addition to the national and international initiatives and support from organizations such as WHO, Rotary International, UNICEF, GAVI to achieve the same goal of complete eradication of polio. National Immunization Days (NIDs) were established in many countries to help immunization of children (Tulchinsky 2018). The global program of polio eradication uses OPV as part of routine infant immunization, including on NIDs. Regions such as the Americas, Europe and China instituted this strategy. However, India still had 42 cases of WPV in 2010, with it's last case ocurring in 2011, being a result of slowness to improve local immunization. In 2012, India was removed from the list of endemic polio and in 2014 certified as polio free by WHO (Tulchinsky 2018).

Complete eradication of polio demands flexibility in vaccination strategies, combining OPV and IPV, aiming for the institution of IPV. The United States uses only IPV on vaccination policy, a strategy adopted by industrialized countries, but difficult to be practiced for developing countries considering the high cost of production of each dose and total doses until they reach herd immunity. IPV-only was instituted in 2000 in several industrialized countries with interruptions of WPV, but, with this decision there is a risk of less intestinal immunization, which raises the chance of introduction of WPV by travelers or refugees (Tulchinsky 2018). Regions which still suffer with enteric disease may need the combination of OPV and IPV, especially in tropical areas where endemic *Poliovirus* and diarrheal diseases are still found. The immunization using OPV required multiple doses to reach the protective antibody levels. Interference on OPV can occur when several enteric viruses are present, facilitating the emergence of VAPP and circulating vaccine-derived polio (cVDPV). The use of IPV as initial choice eliminates the possibility of VAPP (Tulchinsky 2018).

In 2012, World Health Assembly, in concordance with the Director General, started a movement to eradicate polio in the world stipulating the application of health regulations to avoid the reintroduction of WPV from endemic stage nations to neighboring countries. In 2014, WHO recommended that all countries abolish OPV monovalent and introduce bivalent OPV (bOPV - strain 1 and 3) and IPV, which proved to being more effective considering the raising antibody levels, directing the use of IPV as end-game strategy (Tulchinsky 2018).

In this way, the environmental surveillance of sewage and wastewater is an important tool to detect polio. This monitoring program identified the presence of polio in Israel in the year of 2013 and in India in 2012 (Tulchinsky 2018).

Environmental surveillance adds to the epidemiological data, being essential to the post-eradication phase of polio, which has been implemented in several European countries and sub-Saharan African countries. It is still necessary to intensify immunization programs in countries with several problems of sanitation, providing the switch between OPV to IPV available to the population, maintain and implement vigilance environmental programs to identify presence of WPV or VAPP in sewage and wastewater. Therefore, it is fundamental that all the countries adopt and implement these positions to ensure our goal of complete eradication of polio (Tulchinsky 2018).

The *Coxsackieviruses* is a species belonging to the genus *Enterovirus* whose name is a tribute to Coxsackie, a city located in the state of New York, where the virus was first isolated. This species was discovered by Gilbert Dalldorf during a project conducted at the State Laboratories which sought to investigate outbreaks of polio in the state during the summer of 1947 (Tracy et al. 1997, Henry 2012). The *Coxsackievirus* was isolated from the feces of two children who had clinical signs similar to those of poliomyelitis. Using mice as an animal model for viral isolation, Dalldorf realized that the virus was not neutralized by specific antibodies against poliomyelitis, reaching the conclusion that it was a unique species (Dalldorf and Sickles 1948).

By using newborn mice for viral isolation, Dalldorf discovered a new and economically viable animal model for viral recovery, as mice are more susceptible to infections in general at this stage of development (Tracy et al. 1997). Working with newborn mice, Dalldorf and his team were able to distinguish two groups of *Coxsackieviruses*. The first group of viruses was able to lead mice to fatal paralysis through widespread destruction in the animal's skeletal muscle in less than 24 hours. The second group of viruses presents a slower development of the pathology, but with effects that encompass several tissues of the mouse. The first group, responsible for a rapid fatal paralysis by inflammation of the muscles and for a rapid worsening of the health of the studied model, was named *Coxsackievirus A*. The second group of viruses, which affect pancreatic, cardiac, liver, nerve and body cells, but lead to a slower fatal paralysis, were categorized as *Coxsackievirus B* (Gifford and Dalldorf 1951).

After the first work that sought to describe and categorize the *Coxsackievirus* serotypes was already discovered, effort to prepare standardized reagents, antibodies and determine antigens took up a good part of the 1950s. However, a large number of researchers involved with the study of the description of the properties of the *Coxsackieviruses* were becoming infected as a result of incorrect handling of contaminated materials. *Coxsackieviruses* are found in high concentrations both in the tissues of mice used for isolation and in samples obtained from patients, making it essential to establish laboratory standards and techniques that prevent direct contact with these materials. Allied to the large number of researchers who suffered from infection, the immense volume of serotypes being discovered and the drop in research investments in this area after the development of the poliomyelitis vaccine, led the scientific community to turn to the study of oncogenic Enteroviruses, reducing the volume of research focused on the study of *Coxsackievirus* (Tracy et al. 1997).

Coxsackievirus A serotypes, which are currently 23 in number, are mostly causing infections in the back of the throat, a condition known as herpangina (Pozzetto and Gaudin 1999). Between June and July 1957, in the Canadian city of Toronto, 60 people suffered from fever, malaise and soreness in the throat and mouth. Through the efforts of researchers belonging to the School of Hygiene at the University of Toronto, the presence of a viral pathogen with characteristics that included it in the *Coxsackievirus A* group was identified in samples obtained from patients (Robinson et al. 1958).

To determine which *Coxsackievirus A* serotype was responsible for the outbreak, the 18 isolates obtained from the patients were inoculated into monkey kidney cell culture and added with antibodies against different strains of *Coxsackievirus A* and other enteroviruses. The isolates did not have their viral load neutralized by any antibody, with the exception of the antibody against *Coxsackievirus A type 16*. From this result, the research team carried out an experiment inoculating the isolates obtained from the samples of the patients in newborn mice and comparing the histological changes and the effects of the disease manifestation with mice inoculated with standardized viral suspensions of *Coxsackievirus A strain 16*. The disease was identical in all mices, with symptoms that fit the typical clinical picture of *Coxsackievirus A* infections (Robinson et al. 1958).

Hand, foot and mouth disease is a highly contagious condition that usually affects children under five years old, but individuals of any age can be infected. Like other conditions related to *Coxsackievirus A* infections, this disease affects the mouth and throat, causing pain when ingesting water and food, which can generate the symptom of dehydration. Although complications rarely occur, hand, foot, and mouth disease can also cause nail loss, meningitis, encephalitis, and paralysis (NCIRD 2021).

Due to the constant increase in the number of cases of hand, foot and mouth disease that have occurred over the last few decades, this disease has become a public health concern since 1998. In 2008, China included this disease in the list of National Notifiable Diseases Surveillance System. From the time of implementation to October 2014, 11,748,976 cases of hand, foot and mouth disease and 3213 deaths were recorded in the country. More than 90% of cases were related to *Coxsackievirus A* infection (Bian et al. 2015).

The *Coxsackievirus B* group comprises six serotypes that are united because of the unique symptoms that have been described in studies of newborn mice (Pozzetto and Gaudin 1999). In most cases, infections in humans with *Coxsackievirus B* only lead to the development of flu-like symptoms. In some cases, however, the infection can get worse, debilitating the individual for up to two months. The relationship between *Coxsackievirus B* infection and heart disease still needs further studies, but it is possible to affirm the significant role of this group in the case of heart diseases of viral origin (Robinson et al. 1958).

The work by Gaaloul et al. (2014) sought to investigate the role of *Coxsackievirus B* in heart diseases such as myocarditis and pericarditis. Myocarditis is an inflammation of the heart muscle, while pericarditis is a chronic irritation of the pericardium, both conditions being related to comorbidities (Mattingly 1965, Goyle and Walling 2002, Ben-Haim et al. 2009). Myocarditis and pericarditis can present together or individually, but the persistence of this condition can lead to complications. Due to the historical recognition of viral infections as causative agents of heart disease, the research group analyzed the presence of *Coxsackievirus B* genomic RNA in blood and pericardial fluid samples from patients with myocarditis and pericarditis (Gaaloul et al. 2014).

Through molecular techniques involving RT-PCR, it was possible to determine that out of the 102 patients evaluated, 28 tested positive for the presence of enterovirus RNA in blood or pericardial fluid. When performing the genomic sequencing of these samples, it was identified that all had the *Coxsackievirus B* genome, 27 of which had the *Coxsackievirus B type 3* and only 1 sample had the *Coxsackievirus B type 1*. The study was successful in highlighting the significant role of *Coxsackievirus B* in the occurrence of heart disease, being important for the development of new therapies and preventive strategies (Gaaloul et al. 2014).

Echovirus are a group of small, non-enveloped viruses of the *Enterovirus* genus, belonging to the *Picornaviridae* family. They are responsible for causing diseases with symptoms ranging from common colds to fatal meningitis and encephalitis (Haaheim et al. 2002). The group was first accidentally isolated while trying to assess the presence of poliovirus in stool samples from asymptomatic children (Choudhary 2019). After carrying out studies of viral propagation and isolation in cell culture, this group of new viruses were classified as belonging to the same group that contained the *Poliovirus* and *Coxsackievirus*. These species were grouped together due to the presence of similar physicochemical properties, such as resistance to organic solvents and low pH values (Hyypiä and Harvala 2015).

Although these studies demonstrate that these viruses have a specific cytopathic effect when inoculated into cell culture, they were unable to determine the presence of visible pathological lesions, thus making it difficult to study the disease associated with the virus (Choudhary 2019). Due to the properties that were known about this new group of viruses and the lack of association with a clinical picture, these viruses were named ECHO. This acronym symbolizes the fact that they are enteric, as they were isolated from stool samples, cytopathogenic, as they have a cytopathic effect in cell culture, humans, as they did not lead to the development of diseases in other mammals, and orphans, as it was not known which disease pattern they were associated with (Hyypiä and Harvala 2015).

In association with *Polioviruses* and *Coxsackieviruses*, *Echoviruses* became members of the *Enterovirus* genus. Subsequently, changes in the methodology of description and nomenclature of new enteroviruses determined that the new strains isolated would be named after the genus, followed by a number. From that moment on, the species that made up the *Enterovirus* genus were *Poliovirus, Coxsackievirus, Echovirus* and *Enterovirus*. Years later, based on phylogenetic analyses, the International Committee on Taxonomy of Viruses defined that the pre-existing species would have their variants distributed among 12 *Enterovirus* species named *Enterovirus A–L* (Zell et al. 2017). Finally, the genome sequencing of *Rhinovirus* species already described indicated that this group, which previously comprised an exclusive genus, is actually part of the *Enterovirus* genus (King et al. 2011, Palmenberg and Gern 2015).

In addition to other properties that unite the species belonging to the *Enterovirus* genus, such as epidemiology, transmission routes and physicochemical properties, another factor common to *Echoviruses* and other enteroviruses is seasonality. Studies that followed human populations longitudinally over several years indicated that in 90% of cases of *Echovirus* detection in fecal samples, this presence occurred during the period from June to October (Horstmann 1958).

As with other enteric viruses that are transmitted through contact with contaminated water or food, many of the occurrences of *Echovirus* outbreaks have been recorded in communities with medium or low socioeconomic status. Studies carried out in different countries have linked *Echovirus* infection with factors related to precarious socioeconomic conditions such as low annual income. Honig et al. (1956) was one of the pioneers in the epidemiological study of Enteroviruses. Analyzing the excretion of enteric viruses in two communities with different socioeconomic status, a positive correlation was obtained between the lowest income locality and the highest presence of *Echovirus*. Ramos-Alvarez and Sabin (1956) also conducted an experiment evaluating the presence of Enteroviruses in perianal swabs from children from different cities. The vast majority of samples that were positive in cell culture for the presence of non-polio *Enterovirus* belonged to children living in cities with lower economic power (Honig et al. 1956, Ramos-Alvarez and Sabin 1956).

Echovirus species currently has 29 serotypes, making it the highest within the genus. Further, with the inability to replicate in animal hosts, the development of an effective vaccine against this pathogen represents challenges (Choudhary 2019). As the principle of protection related to vaccine development is based on the use of specific antibodies to neutralize a given serotype, the existence of a large variety of serotypes makes it difficult to establish a single vaccine that is effective for all (Hyypiä and Harvala 2015).

The classification and nomenclature of enteroviruses are constantly evolving and the species of this heterogeneous group are currently characterized mainly based on the homology of the nucleotide sequences of RNA molecules. The existence of *Enterovirus* species named only with the genus and subsequent numbers is related to changes in the classification and nomenclature pattern that occurred during the discovery of new species (ICTV 2021). From 1960 onwards, the new *Enterovirus* species that were isolated, no longer received names such as *Echovirus, Poliovirus* and *Coxsackievirus* and started to be named through consecutive numbers, as in *Enterovirus 68* (Simmonds et al. 2020).

Over the years, phylogenetic analyzes have shown that within groups united on the basis of their pathogenesis, such as *Coxsackievirus A*, grouped due to their ability to rapidly lead to paralysis, showed great genetic heterogeneity. As a solution, the existence of four enteroviruses species was determined: *Enterovirus A, Enterovirus B, Enterovirus C* and *Enterovirus D*, based on the genetic similarity between the strains included in each group. *Polioviruses*, with their capacity to cause severe paralytic disease and neurological damage, belong to the *Enterovirus C* species. *Coxsackievirus A* had their strains distributed among the *Enterovirus A, Enterovirus B* and *Enterovirus C* species. The *Rhinovirus* species are an exception, since they have kept this name (Simmonds et al. 2020). The current classification defined by ICTV defines the existence of 15 species within the *Enterovirus* genus, 12 of which are named *Enterovirus A–L* and 3 species of *Rhinovirus* (*A, B* and *C*) (ICTV 2021).

This new classification can be confusing. Since it is based on similarity between the nucleotide sequences of the genome of *Enterovirus* species, groups that have properties in common, such as tropism, pathogenicity and resistance to unfavorable conditions are now separated into different species. Similarly, strains that do not have properties in common are classified as belonging to the same species. Disagreements still occur with regard to this nomenclature model, since the similarity between the genomic sequence is the only factor taken into account, disregarding the properties that characterize and distinguish the enteroviruses (Bessaud et al. 2018).

The work elaborated by Oberste et al. (2005) sought to identify the *Enterovirus* strains present in stool samples from patients with gastroenteritis. Through the genetic sequencing of these samples, it was demonstrated that they had < 70% similarity between the analyzed genomes and the previously described *Enterovirus* serotypes, suggesting that they were new strains. Phylogenetic analysis of these samples indicated a similarity of 54.9–69.4% with the genomes of strains belonging to the species *Enterovirus A*. The new strains discovered were then named according to the current nomenclature standard: *Enterovirus A76, Enterovirus A89, Enterovirus A90* and *Enterovirus A91* (Oberste et al. 2005).

The *Rhinovirus* group had its first representative isolated in 1956 by Dr. Winston Price at Johns Hopkins University. Nasopharyngeal lavage samples from patients with respiratory conditions were inoculated into cell culture and their cytopathic effects were described (Price 1956). In addition to the common flu symptoms witnessed by patients who had their samples analyzed by Dr. Price, such as fever, coryza and malaise, *Rhinovirus* infection can also affect the lower respiratory tract and aggravate chronic respiratory conditions such as asthma (Kennedy et al. 2012). Subsequently, the new isolated *Rhinovirus* variants were classified into *Rhinovirus A* and *Rhinovirus B* according to their aminoacidic sequence (Laine et al. 2005).

Currently, the *Rhinovirus* species belongs to the *Enterovirus* genus, but this was not always the case. In 2008, members of the International Committee on Taxonomy of Viruses chose to make changes to the current classification. Among the proposals that were validated is the inclusion of the *Human Rhinovirus A* and *Human Rhinovirus B* species in the *Enterovirus* genus and the elimination of the *Rhinovirus* genus that harbored these species within the *Picornaviridae* family (Carstens and Ball 2009). The *Enterovirus* genus was then described as containing 4 *Enterovirus* species (*A, B, C* and *D*), which include the *Poliovirus, Echovirus* and *Coxsackievirus* species, and two *Rhinovirus* species (*A* and *B*) (ICTV 2009).

The reason for the union of the *Enterovirus* and *Rhinovirus* genera was the several structural and genetic similarities presented by the species that compose them. *Rhinovirus* and *Enterovirus* share the same virion structure, have a virtually identical genome expression strategy and have great similarity between genome organization and nucleotide sequence (Laine et al. 2005). The main differences, which ended up being responsible for the previous separation of these species between two distinct genera, are tropism and resistance. While enteroviruses replicate primarily in the gastrointestinal tract, with some representatives also infecting the respiratory tract, most rhinoviruses are restricted to the respiratory mucosa. *Enterovirus* are also able to withstand varying conditions of temperature and pH, while *Rhinovirus* are more susceptible to being inactivated by these factors (Laine et al. 2005, Simmonds et al. 2010).

Since 2007, studies on a divergent variety of rhinoviruses have led to genetic sequencing and phylogenetic analyses. The results pointed to the existence of a third species of *Rhinovirus*, named *Rhinovirus C*. The three species of *Rhinovirus* generate infections with similar symptoms, affecting the upper respiratory tract and aggravating chronic respiratory conditions such as asthma. *Rhinovirus C*, however, has been indicated as responsible for lower respiratory tract infections and aggravating asthma in adults (Lau et al. 2007).

As one of the main respiratory pathogens causing influenza, *Rhinoviruses* impact society due to the number of school and work hours that are lost annually due to *Rhinovirus* infections. It is estimated that about 50% of common flu occurrences are the result of infection with *Rhinovirus* variants. Although it does not generate serious clinical conditions, there are economic and social impacts that take on alarming proportions when the infection ends up aggravating chronic respiratory syndromes such as asthma (Turner 2007).

1.2 Main species

The genus *Enterovirus* includes 15 species with about 175 different serotypes, of which *Enteroviruses A–L* are oral-fecal transmitted and *Rhinoviruses A–C* are respiratory ones (ICTV 2021). Polyhedral and non-enveloped, they have a linear ssRNA(+) genome of approximately 7.2–8.5 kb and a icosahedral capsid of approximately 30 nm consisting of 60 subunits with 3 structural proteins (VP1, VP2, VP3), VP4 is located inside the capsid (Harvala et al. 2018, Casas et al. 2001). Enteroviruses are the most common group of pathogens which affect humans and have a high infant mortality rate in children under 1 year-old (Palacios et al. 2005). Its former taxonomic classification was based on the most diverse clinical manifestations, from the mildest, such as cold and flu, to the most severe such as diarrhea, or even encephalitis/paralysis and meningitis. Nowadays, with the advancement of genomic analysis techniques, the new classification is based on the similarity of the nucleotide sequence and its organizational order (ICTV 2021). Thus, there are 5 main groups that will be better discussed in this chapter: *Enterovirus, Poliovirus, Coxsackievirus, Echovirus* and *Rhinovirus*.

Among enteroviruses there are several serotypes that have different temporal patterns of circulation and are often associated with different clinical manifestations.

Changes in circulating serotypes can be accompanied by large-scale outbreaks. Beginning in the 1960s, newly discovered enteroviruses are given a numerical designation (e.g., *Enterovirus 71*) rather than being assigned to one of the traditional groups (Khetsuriani et al. 2006).

The current taxonomy takes into account molecular and biological features and divides human enteroviruses (HEV) into twelve species (*Enterovirus A–L*), but retains traditional names for individual serotypes. With molecular techniques of *Enterovirus* typing, new HEVs continue to be identified, in particular the *Enterovirus 79–101* that have been described recently. As an example we have *Enterovirus 71* (HEV-71) which is associated with severe neurological manifestations, mild skin rashes and hand, foot and mouth disease. During the 1960s and 1970s, several outbreaks of paralysis were reported in Europe and linked to this *Enterovirus* serotype (Khetsuriani et al. 2006, Ooi et al. 2010).

Another interesting case is *Enterovirus 68* (HEV-68) which is the only one of its subgroups that has rhinovirus properties as well. However, HEV-68 is one of the most rarely reported serotypes. Usually *Enterovirus 68* is detected in respiratory samples and in a single known case of acute flaccid paralysis, in 2005. This case was the first report of *Enterovirus 68* central nervous system infection in a young adult. The most commonly affected age group is children between 1 to 4 years old (Khetsuriani et al. 2006).

The *Poliovirus* is a member of the genus *Enterovirus* of the *Picornaviridae* family, which causes asymptomatic infection in the human intestine. It can destroy parts of the nervous system, leading to weakness, and can cause meningitis, encephalitis, myelitis, and muscular paralysis, generating poliomyelitis. Poliomyelitis is a very old human disease which became a major health problem in the early 20th century. This virus infects human cells through the CD115 receptor found on both humans and some primate, which is an integral membrane protein and member of the immunoglobulin superfamily. Infection usually entails replication of the virus within the gastrointestinal tract and subsequent excretion in the feces (Minor 2021).

The *Coxsackievirus* is also a member of the *Enterovirus* genus of the *Picornaviridae* family, but it is separated into two groups, Group A and Group B. *Coxsackievirus* group *A* (1–22 and 24) is composed of 23 subtypes that infect skin, mouth, nails, and eyes, and Group B (types 1–6) has 6 subtypes that infect the heart, pleura, pancreas, and liver. Its means of transmission is through direct contact with infected people, usually by poorly washed hands contaminated with feces, with group A causing hand, foot, and mouth disease. The CD55 protein functions as a receptor for certain coxsackieviruses, however it is not sufficient, needing ICAM-1 molecules or α- and β-integrins as correctors (Di Prinzio 2022).

The nomenclature of *Echovirus* is derived from *Orphan Human Cytopathic Enteric* where orphan refers to viruses that are not associated with any disease but may carry pathogenicity (Li and Delwart 2011). Thus, initially many viruses were referred to the group *Echovirus*, but currently there are only 29 serotypes classified as such (Choudhary 2019). The *Echovirus* penetrates the cells of the pharynx and intestine and reaches the M cells of the local lymph nodes, causing subclinical viremia. From there, it spreads to reticuloendothelial tissues such as the liver, spleen, and bone marrow, and to the other lymph nodes. Currently, there are at least 2 cellular receptors recognized by echoviruses, depending on their serotype: the $\alpha 2$ subunit of the VLA-2 integrin (E-1 and E-8) and decay accelerating factor CD55 (E-6, E-7, E-11, E-12, E-20, E-21, E-29, and E-33) (Bergelson et al. 1993a, Bergelson et al. 1994b, Hyypiä and Harvala 2015). Although over 90% of infections are asymptomatic, it often causes a general/febrile malaise and more rarely acute aseptic meningitis, encephalitis, pericarditis or paralysis (Yin-Murphy and Almond 1996, Romero and Modlin 2014). Recently, a novel neonatal receptor Fc (FcRn) present in intestinal enterocytes, hepatocytes and the endothelial cells of the blood-brain barrier of newborns has been discovered, which explains their increased susceptibility and greater severity of *Echovirus* infections. Viral excretion is perpetuated up to 5–6 weeks after infection and treatment is based on antivirals (Morosky et al. 2019).

Rhinovirus, as their name suggests (*rhinos* from Ancient Greek is "from the nose"), are the only subgroup of enteroviruses that primarily target the respiratory tract. Rhinoviruses become unstable at pH below 5–6 and on surfaces outside the host after 3–4 hours. They initially infect the epithelium of the upper respiratory tract such as the nose, where lower temperatures (32–35°C) favor their replication (Kennedy et al. 2012). Their main routes of transmission are aerosols, direct contact with infected people or fomites (Yin-Murphy and Almond 1996). Rhinoviruses can be separated into three groups according to their target receptor: the major-group that includes most *Rhinovirus A* and all *Rhinovirus B* and affects intercellular adhesion molecule 1 (ICAM-1), the minor-group that includes 12 *Rhinovirus A* serotypes and affects low-density lipoprotein receptor (LDLR), and the newer group that harbors *Rhinovirus C* and affects the cadherin related family member 3 receptor (CDHR3) (Schuler et al. 2014, Blaas and Fuschs 2016, Husby et al. 2017, Basnet et al. 2019). They are responsible for more than 50% of cold cases in adults, which usually resolve within 14 days, although they can present complications such as bronchitis and pneumonia (Husby et al. 2017). Treatment is restricted to the administration of symptom-relieving drugs, rest, and the prevention of person-to-person transmission (Yin-Murphy and Almond 1996).

1.3 Pathogenesis of enteroviruses

Enterovirus infections are more frequent in children under the age of 10. Although most infections with enteroviruses are asymptomatic, some can be severe and life-threatening. *Enterovirus* transmission occurs via the fecal-oral route or by respiratory droplets and thereby it is strongly associated with poor sanitation (Muehlenbachs et al. 2014).

The infection with enteroviruses has a wide range of clinical presentations, including neurological diseases (*Poliovirus*, *Enterovirus-A71* and *Coxsackievirus A*), meningitis and meningoencephalitis (*Coxsackievirus A*), myocarditis and pericarditis (*Coxsackievirus B* and *Echovirus*), pancreatitis (*Coxsackievirus B* viruses), hand, foot and mouth disease (*Enterovirus-A71* and *Coxsackievirus A*) and respiratory disease (*Rhinovirus* and *Enterovirus-D68*) (Muehlenbachs et al. 2014).

Once they are ingested, they infect cells of the oropharynx and intestine. Then, they replicate in the regional lymph nodes and are released into the bloodstream.

They may reach different locations such as the heart, pancreas, and/or the respiratory tract causing, respectively, myocarditis, pancreatitis, and respiratory illnesses (Romero 2017). Enteroviruses can also invade the central nervous system (CNS), causing meningitis, encephalitis and acute flaccid paralysis (Majer et al. 2020, Elrick et al. 2021).

Three distinct mechanisms of neuroinvasion have been reported: retrograde axonal transport, infection of circulating immune cells, and the direct crossing of the blood-brain barrier. *Poliovirus*, *Enterovirus-A71* and *Enterovirus-D68* use the retrograde axonal transport within motor neurons to enter the CNS (Elrick et al. 2021). They enter the cells via receptor-mediated endocytosis and travel throughout the cell in the interior of endosomes. The *Enterovirus-A71* was shown to be able to replicate in CD14+ cells, dendritic cells, and peripheral blood mononuclear cells (PBMCs) and therefore may use these cells as viral shuttles to reach the CNS. *Coxsackievirus B3* and *Echovirus 40* can directly infect the blood-cerebrospinal fluid barrier, which allows the direct crossing of the blood-brain barrier (Majer et al. 2020, Sinclair and Omar 2022).

Within the nervous system, polioviruses and *Enterovirus-D68* damage motor neurons of the anterior horns in the spinal cord. The death of these cells leads to irreversible paralysis. In contrast, *Enterovirus-A71* presents tropism for the brainstem, pons, medulla, cerebellum, cortex, thalamus, dentate nuclei, and cerebrum. Enteroviruses cause damage within the CNS by inducing pyroptosis, apoptosis, or autophagy pathways. Those mechanisms can be triggered either by viral replication or the attempt of the host immune system to minimize the damage caused by viral infections (Majer et al. 2020, Sinclair and Omar 2022, Wiley 2020). Unlike *Poliovirus*, there is no vaccine available for *Enterovirus-D68* and *Enterovirus-A71* (Sooksawasdi Na Ayudhya et al. 2021).

Rhinoviruses infect cells from the epithelium of the respiratory tract. These viruses enter the cells via endocytosis or pinocytosis. The infection caused by rhinoviruses does not result in direct cell destruction. They disrupt the epithelial barriers by stimulating the production of reactive oxygen species (ROS) and dissociation of proteins from the tight junctions. In addition, the infection by rhinoviruses triggers the release of cytokines that activate immune cells, such as granulocytes, dendritic cells, and monocytes (Sinclair and Omar 2022).

Enterovirus infections can also cause myocardial inflammation, myocarditis, which is most often self-limited and subclinical. *Enterovirus* myocarditis is rare, affecting mainly adults between 20 and 40 years of age, being more severe in neonatal patients. In these cases, *Coxsackievirus* group *B* and *Echovirus* are the most frequently identified. These viruses bind to the *Coxsackievirus* and the *Adenovirus* receptor (CAR) and infect cardiomyocytes, which can trigger myocardial injury (Muehlenbachs et al. 2014, Romanos et al. 2015, Tschöpe et al. 2021).

Hand, foot and mouth disease is caused by strains of coxsackieviruses, but in most cases by *Coxsackievirus A type 16* and *Enterovirus-A71*. It most often affects young children. The Scavenger receptor class B, member 2 (SCARB2) is the receptor for *Enterovirus-A71* and *Coxsackievirus A type 16*, and is reported in squamous cells of the tonsillar crypts, in addition to studies reporting squamous epitheliotropism of *Enterovirus-A71* in the epidermis and oral mucosa. These sites are the foci of viral replication and sites of clinical manifestation of the disease (Phyu et al. 2017, Muehlenbachs et al. 2014).

Enteroviruses can be detected in oral secretions, rectal swabs, and stool. In stool samples, they can be detected months after the symptoms resolve (Sinclair and Omar 2022). Although a prior exposure to an *Enterovirus* does not prevent re-infection, during re-infection viral replication is limited to the site of primary replication (i.e., gastrointestinal and respiratory epithelium), without viral dissemination to other tissues (Muehlenbachs et al. 2014).

1.4 Detection in the environment

Enteroviruses are pathogens transmitted mainly via the fecal-oral route, that is, through contact with contaminated saliva, water or food. *Echovirus*, *Poliovirus* and *Coxsackievirus A* and *B* are examples of microorganisms belonging to the Enterovirus genus that are transmitted through these routes. They are excreted in high concentrations in the feces of infected individuals, beyond that, their persistence for long periods in unfavorable environmental conditions contribute to numerous outbreaks of gastroenteric diseases, meningitis and encephalitis worldwide. Due to inappropriate waste disposal and the possible ineffectiveness of routine water treatment systems, residual pathogens end up contaminating the environment and reaching water resources, compromising public health (Fong and Lipp 2005).

Infected individuals can excrete *Enterovirus* (EV) particles in feces for up to 16 weeks and in concentrations that can reach 10^{11} viral particles/gram of feces (Romero 1999, Fong and Lipp 2005). Once in the environment, EVs are able to tolerate unfavorable conditions of pH, temperature and salinity, remaining viable for long periods of time (Rajtar et al. 2008). In addition to these properties, the genus is made up of non-enveloped viral particles, which guarantees greater resistance to disinfection processes (Pellegrinelli et al. 2013). Due to this ability to contaminate different types of environment, to remain infectious for a long period of time and because they have a low infectious dose, EVs become a public health concern (Connell et al. 2012).

In 2007, Law 11.445 was published, making the State responsible for expanding basic sanitation in the country (Presidência da República 2007). The last IBGE census, carried out in 2019, indicated that only 68.3% of Brazilian households have access to sanitary sewage (IBGE 2019). Ideally, the wastewater that composes the sewage should be collected and sent to the Wastewater Treatment Plant (WWTP) responsible for disinfection and subsequent return of these waters to the environment. However, the common treatment, routinely carried out in WWTPs, is unable to completely inactivate the pathogens in the sewage (Connell et al. 2012). In addition to the inefficiency of routine sewage treatment, a large portion of the Brazilian population still does not have access to basic sanitation and worldwide 4.2 billion people lack safely managed sanitation services, with 80% of the wastewater produced being disposed of without any treatment back into the environment (UNESCO 2017, WHO 2019).

The outbreaks of viral gastroenteritis reported worldwide are related to water pollution from wastewater composed of viral particles released into surface water (Assis et al. 2018, Barril et al. 2015). In fact, EVs have already been isolated from wastewater, rivers, marine water, ponds, groundwater, potable water and recreational water (Pianetti et al. 2000, Borchardt et al. 2003, Costán-Longares et al. 2008, Cesari et al. 2010, Okoh et al. 2010). The use of insufficiently treated wastewater to irrigate fruit and vegetable crops also poses a risk to human health, since there is the possibility of infection by viral particles when these products are eaten (Steele and Odumeru 2004). Table 1 provides examples of cases in which species of the *Enterovirus* genus were detected in environmental samples.

Despite the large volume of contaminated wastewater that is improperly discharged into the environment, microbiological water quality standards are still, in most countries, verified based only on the presence of bacterial contamination. The use of bacterial indicators such as fecal *Enterococci*, Fecal and Total Coliforms as indicative of environmental contamination is not an efficient method, as there is no scientific evidence that proves the correlation between the presence of bacteria and viral contamination (Schvoerer et al. 2001). Differences in size, physical properties and resistance to unfavorable environmental conditions ensure that the virus persists in its infectious form for longer when compared to bacterial contamination (Wurtzer

Table 1. Examples of cases of Enterovirus detection in environmental samples.

Location and date	Type of sample	Investigated viruses	Reference
Rome, Italy (1997)	Recreational water	*Echovirus 30*	Faustini et al. (2006)
Pretoria, South Africa (2000–2002)	Wastewater, river water, drinking water and spring water	*Coxsackievirus A and B* and *Echovirus*	Ehlers et al. (2005)
Barcelona and Girona, Spain (2001–2006)	Wastewater, seawater and river water	*Coxsackievirus A and B, Echovirus* and *Poliovirus*	Costán-Longares et al. (2008)
Mexico (2004)	Sea water	*Echovirus 30* and *Coxsackievirus A*	Begier et al. (2008)
Milan, Italy (2006–2010)	Wastewater	*Coxsackievirus B, Echovirus* and *Enterovirus*	Pellegrinelli et al. (2013)
Bolzano, Naples and Palermo, Italy (2007–2010)	Wastewater	*Coxsackievirus B and Echovirus*	Battistone et al. (2013)
Hawaii, United States of America (2010–2011)	Seawater and freshwater	*Coxsackievirus A and B, Echovirus* and *Enterovirus 68*	Connell et al. (2012)
Florianópolis, Brazil (2011)	Freshwater and drinking water	*Enterovirus*	Nascimento (2011)
Apulia, Italy (2014)	Groundwater	*Enterovirus*	De Giglio et al. (2017)
Potters Bar, England (2021)	Wastewater	*Enterovirus 68*	Tedcastle et al. (2022)

et al. 2014). EVs are even resistant to the chlorine treatment commonly used in most WWTPs, facilitating their survival in treated sewage (Gregory et al. 2006). Due to their high concentration in the excreted feces, great variation in pH, temperature and salinity that they are able to resist, the long period which they remain viable under unfavorable conditions and the high number of individuals affected annually by viral gastroenteritis, Enteroviruses have been recognized as potential indicators of environmental contamination (De Giglio et al. 2017).

Despite the importance of using alternative indicators, such as EVs, most countries are still under legislation that assesses the microbiological standard of water quality based exclusively on the presence of bacterial indicators. In Brazil, the Ministry of Health published Ordinance No. 888 in May 2021 that replaced the water potability standards defined in Annex XX of Consolidation Ordinance No. 05 of 2017 (Ministério da Saúde 2021). This update continued to exclude the obligation to analyze the presence of enteric viruses in the potability standard, with microbiological quality still being evaluated exclusively considering the presence of bacterial contamination. The search for the presence of viral contamination is only recommended in situations of acute diarrheal disease outbreaks. In addition to ignoring the necessity to assess the presence of these pathogens responsible for several outbreaks of non-bacterial gastroenteritis, the ordinance was also responsible for eliminating the need for analysis of some chemical parameters and increasing the tolerable limit for others (Ministério da Saúde 2021).

The constant cases of outbreaks of pathologies related to contact with water contaminated with viral agents are a stimulus for the development of more effective techniques for water treatment, but it is still essential to establish continuous monitoring systems for water quality. EVs have already been isolated in a wide variety of environmental samples of different characteristics, representing a risk for individuals who use water resources for food or recreational activities, since water becomes an efficient vehicle for pathogen transmission (Rajtar et al. 2008).

Despite the obvious importance of assessing viral contamination in the environment, the diagnostic methods employed still face challenges in terms of virus recovery. The viral isolation process, which can be long and laborious, depends on the sample processing technique, since the different properties they present make it necessary to apply specific analysis methods (Haramoto et al. 2018). Even considering the large concentration of viruses excreted in feces and dumped in the environment, the number of viral particles is still small when compared to the volume of water that a sea, river or lagoon represents, for example. To solve this question, the sample processing for the diagnosis of the presence of EVs in environmental samples necessarily includes a concentration step whose objective is to reduce the volume of the sample, optimizing the detection assays (Katayama et al. 2002, WHO 2003, Ikner et al. 2012, Cashdollar and Wymer 2013).

Depending on the sample properties, this concentration step will differ. Some of the concentration techniques already applied in the analysis of environmental samples are adsorption-elution, coagulation/flocculation, two-phase separation using polyethylene glycol, ultracentrifugation, precipitation by addition of salt, among others (Haramoto et al. 2018). The development and application of these concentration methods have considerably increased the effectiveness of both molecular and cell culture detection techniques (Ikner et al. 2012, Cashdollar and Wymer 2013).

In freshwater, river water, drinking water, groundwater, tap water and surface water samples, the concentration technique based on elution followed by adsorption is the most effective. Katayama et al. (2002) was responsible for describing this technique when searching for the most effective method to assess the presence of enteric viruses in large volumes of water. In this concentration model, a predetermined volume of sample is made positive through the addition of a positively charged ionic solution. The sample is then filtered in a vacuum pump system through a negatively charged membrane. Due to the difference between the charges of the sample and the membrane, the viruses are adsorbed to the surface of the membrane and subsequently washed with a buffer solution for viral recovery. This concentrated solution can then be used in detection assays (Katayama et al. 2002, Haramoto et al. 2009, Victoria et al. 2009).

This concentration model has the advantage of being able to adapt itself according to the properties of the water being analyzed. The volume of water corresponding to the required viral concentration will vary depending on the source. While volumes of 40 L are used to perform viral concentration in river water samples, 1000 L are recommended for tap water evaluation (Hata et al. 2015). Another advantage is that the elution step is performed using a simple pH-controlled buffer, which reduces the inhibition often related to the use of organic eluents in molecular assays such as RT-qPCR (Haramoto et al. 2018).

In marine water samples, the method that has shown greater efficiency is flocculation using acidified organic milk. This simple and easy method starts with the process of acidifying the sample and then adding an acidified milk solution. Due to the tendency of the viruses to aggregate to solid particles that protect them from inactivating factors, the adsorption and subsequent decantation of the virus with the milk flakes occurs (Gerba and Schaiberger 1975, Templeton et al. 2008). After eight hours of sedimentation, the supernatant is discarded and the decanted material is collected, centrifuged and the pellet resuspended with saline buffer. This viral suspension is used in detection assays (Calgua et al. 2008).

This methodology has been gaining recognition due to its practicality, low cost and efficiency in viral recovery, especially in marine water. The efficiency that the organic flocculation technique has in contrast to adsorption-elution method when applied to marine water is due to the salinity, conductivity and concentration of cations in the sample. In order for the virus to adhere to the milk particle, a minimum conductivity value in the sample is required. Freshwater samples do not reach this threshold, requiring the addition of sea salt for the flocculation technique to obtain good viral recovery results (Calgua et al. 2013).

Regarding wastewater samples, the concentration technique usually employed is based on the use of polyethylene glycol (PEG). In this method, a smaller volume of sample is used, since the number of viral genomic copies present in wastewater is usually greater than that present in water samples from lakes or rivers, for example. While the adsorption-elution method described by Katayama et al. (2002) suggests the use of a volume of two liters and the organic flocculation technique requires the collection of ten liters of sample, for the concentration method using PEG only 100–200 mL of wastewater sample is necessary (Farkas et al. 2021).

After adjusting the pH, the PEG-containing solution is added to the samples, which are then kept under agitation to enable the viruses to bind to the PEG particles. Afterwards, the samples are centrifuged in order to form the pellet. This pellet is then resuspended and this viral suspension can be used in detection assays (Farkas et al. 2021). In the protocol development work carried out by Farkas and his team, the efficiency of the wastewater concentration technique using PEG was tested in relation to the recovery of SARS-CoV-2. A solution of known concentration from the *Murine Coronavirus* (MNV) model was added to the wastewater samples and after the concentration step, RT-qPCR was performed to determine the percentage of virus that had been recovered. The MNV recovery percentages between 9–37% indicate that despite there being a loss in the concentration of existing viruses, it is an efficient technique with interesting applications (Farkas et al. 2021).

In most parts of the world, the wastewater is treated by the activated sludge process, resulting in sludge and biosolids formation (Pepper et al. 2006, National Research Council 2002).

The sludge can be divided into primary and secondary. The primary sludge results from the settling of solids as they enter a wastewater treatment plant. The secondary sludge results from the conversion of soluble organic matter in the wastewater to bacterial biomass (Corpuz et al. 2020, Mohapatra et al. 2021, Bibby and Peccia 2013, Pepper et al. 2006, Yin et al. 2018). The combination of the primary and secondary sludge produces the sewage sludge, and it must be disposed of or recycled in some manner (Pepper et al. 2006). Insufficient sludge treatment may result in viral particle dissemination into the environment since the viruses can adhere to solid particles in wastewater treatment processes and end up as sewage sludge (Gholipour et al. 2022).

Biosolids are widely used in agriculture and non-agriculture for sludge management by improving the physical and chemical properties of the soil (Pepper et al. 2006). It can be classified into two categories according to its microbiological quality. Class A biosolids result from a high-level sludge treatment and considerable levels of pathogens are not found. Class B biosolids result from low-level sludge treatment and considerable levels of pathogens, such as bacteria, parasites, and viruses, are encountered (United States Environmental Protection Agency 2020).

Since developing countries often produce class B biosolids, the application of these biosolids can pose a risk to the health of individuals due to the presence of pathogenic microorganisms, especially enteric viruses. Given this scenario, the development of methods for monitoring and detection of enteric viruses can help to prevent and reduce the potential risks associated with human health.

Cell culture is known as the gold method for examining the infectivity of isolated viruses (Metcalf et al. 1995, Monpoeho et al. 2001). Researchers conducted environmental surveillance of poliovirus from 2006 to 2010 with sewage samples collected from four treatment plants using RD and L20B cells. The *Poliovirus* serotype was identified through neutralization tests with monospecific anti-PV pooled sera. The characterization of the isolated poliovirus as wild or Sabin was performed at the ISS using the ELISA (enzyme-linked immunosorbent assay) test and the PCR (polymerase chain reaction) kit. The identification of the non-polio enteroviruses (NPEVs) isolated was carried out by serum neutralization with pooled sera against NPEV. *Echovirus* represented 58% of Human Enteroviruses serotypes detected in all the wastewater treatment plants, followed by 26% of *Coxsackievirus* group *B*, 12% of enteroviruses mixture, 3% of not typed enteroviruses and 1% of *Poliovirus* (Pellegrinelli et al. 2013).

Although the cell culture is standard for viral isolation, this technique is laborious, time-consuming, often underestimates the level of virus, and cannot differentiate between different serotypes present in a single sample. In addition, since enteric viruses can establish infection in humans at low infectious doses, extremely sensitive detection assays are needed (Schwab et al. 1996, Larivé et al. 2021). Therefore, in many cases, viruses are detected using PCR (Schlindwein et al. 2010, Battistone et al. 2013).

Researchers developed a new real-time reverse transcriptase PCR (RT-qPCR) in-house for *Poliovirus 1* and *Coxsackievirus B type 3* detection in environmental water. And compared its performance with four commercial RT-qPCR kits for enterovirus detection. The results showed that the in-house method provided the best performance of qualitative detection. Since the low inhibition rate permitted enterovirus genomes quantification (Wurtzer et al. 2014).

In another study, researchers searching for an effective laboratory protocol for enterovirus detection developed a highly sensitive and optimized reverse transcriptase PCR (RT-PCR). They compared eighteen published enterovirus primer pairs for detection sensitivity. The primer set exhibiting the lowest detection limit under optimized conditions was validated in a field survey of 22 recreational bodies of water located around the island of Oahu, Hawaii. Since PCR can't differentiate between infectious and non-infectious particles (Boehm et al. 2003), samples positive for enterovirus by PCR amplification were tested in an initial infectivity assay by infecting buffalo green monkey kidney (BGMK) and A549 cell lines (Connell et al. 2012).

Another study evaluated the presence and seasonal distribution of polio and other enteroviruses in four wastewater treatment plants in three cities in Italy, using different treatment systems at both inlets and outlets of the treatment plants. Viral serotypes isolated before and after water treatment were also compared. The researchers led experiments on enteroviruses detection in cell cultures. Then, they perform RT-PCR of the positive samples (with cytopathic effect) to confirm the presence of enteroviruses. Forty-eight non-polio enteroviruses were isolated at the inlet of the four wastewater treatment plants, 35 of which were *Coxsackievirus B* (72.9%) and 13 *Echovirus* (27.1%). After treatment, two *Coxsackievirus B type 3,* one *Coxsackievirus B type 5*, and one *Echovirus type 6* were isolated. *Coxsackievirus B type 3* and *Echovirus type 6* serotypes were also detected in samples collected at the inlet of the treatment plants, in the same month and year (Battistone et al. 2013).

Since both techniques presented have limitations, some studies use integrated cell culture followed by PCR (ICC-PCR). This technique combines cell culture and PCR to quantify enterovirus infectivity in a sample containing multiple species and serotypes and can assess cultural viability.

To quantify infectious enteroviruses polluting the coastal seawaters, a study developed integrated cell culture and reverse transcription-quantitative PCR (ICC-RT-qPCR) at Bohai Bay, Tianjin, China (Ming et al. 2011). Another study used the same approach to quantify the infectious concentrations of eight enteroviruses serotypes commonly encountered in sewage (*Coxsackieviruses A type 9*, *Coxsackievirus B types 1, 2, 3, 4* and *5*, and *Echovirus type 25* and *30*). The method used two cell lines for virus replication and serotype-specific real-time PCR (qPCR) primers for quantification (Larivé et al. 2021).

Nowadays, with the advancement in molecular techniques, different studies have employed next-generation sequencing techniques to achieve an in-depth view of the real diversity of each sample such as diversity, seasonality, and emerging types of enteroviruses in environmental samples. Researchers achieved whole-genome sequencing of enteroviruses, enabling the specific strains identification of *Enterovirus A*, *Enterovirus B*, *Enterovirus C* and *Enterovirus D* present in samples during a single sequencing run (Joffret et al. 2018). In another study, the next-generation sequencing approach was used with reverse transcriptase-polymerase chain reaction products synthesized directly from sewage concentrates. In this survey, the results obtained were capable of determining whole-capsid genome sequences of multiple enterovirus strains from all 4 A to D species present in environmental samples (Majumdar et al. 2018).

This methodology constitutes a great contribution to the knowledge about the circulation, diversity, and patterns of enteroviruses in populations and the data obtained in these studies can be used to strengthen environmental surveillance.

1.5 *Relevance of environmental investigation*

Through the development of our society, the urban centers organize themselves with the help of public measures. The established economic model advocated an uneven distribution reaching nations and the population. In this economic race some countries were able to accumulate more resources than others (Armstrong 2011). Beyond that, the natural resources have been noticed carefully in the last decade, because they are exhaustible and several pathogens can use them to conclude their infection. Hydric resources are essential to every population, used in food production, to hydrate animals and humans and for industrial purposes (Schyns et al. 2019).

The management of water as an essential resource must involve delicate planning, so that the collection and treatment of sewage is efficient in order to obtain an effluent that does not generate risks to the environment. The treatment involves removal of organic particles and drastically decreases pathogens. Undeveloped countries statistically have poor sanitary conditions in comparison with developed countries, which result in investment in public policies that guarantee acceptable quality of life. Countries without satisfactory sanitary conditions allow that pathogens which use the water route to easily infect their population.

Several enteric viruses use this route, resulting in contamination of lakes, groundwater, food and other waterbodies (Fongaro et al. 2015, Bouseettine et al. 2020, Cadamuro et al. 2021). The contamination of water does not only reach the sphere of human health, but generates contamination of soil, environment and consequently animals/food production. The concept of One Health aims to mitigate and act in all spheres, stopping a chain of contaminants that involves several parts of the whole scenario. The eradication of waterborne viruses cannot be resumed in treatment of sewage, but it needs to be visualized as one piece, requiring the fitting of other parts.

Environmental surveillance cooperates with data obtained by epidemiology, gaining visibility and importance as a tool to identify the presence of viruses in the environment. Recently the pandemic of SARS-CoV-2 was related to sewage samples being excreted by infected humans (Prado et al. 2020, Fongaro et al. 2021, Gawlik et al. 2021). Sentinel programs were implemented to identify possible outbreaks before symptoms and to relate and complement epidemiology data (WHO 2003). Therefore, environmental surveillance is a necessary tool to identify the presence of WPV and VAPP in countries, being a measure of failure on immunization programs and/or failure of sanitary conditions, demanding public policies (Asghar et al. 2014). The eradication of polio and other viruses, require improvements on several policies of a country and collaboration as whole.

Reference

Armstrong, C. 2011. Global Resource Distribution. Encyclopedia of Global Justice, 441–443.

Asghar, H., Diop, O.M., Weldegebriel, G., Malik, F., Shetty, S., El Bassioni, L. et al. 2014. Environmental surveillance for polioviruses in the global polio eradication initiative. The Journal of Infectious Diseases 210(suppl_1): S294–S303.

Assis, A.S.F., Fumian, T.M., Miagostovich, M.P., Drumond, B.P. and da Rosa e Silva, M.L. 2018. Adenovirus and rotavirus recovery from a treated effluent through an optimized skimmed-milk flocculation method. Environmental Science and Pollution Research 25(17): 17025–17032.

Barril, P.A., Fumian, T.M., Prez, V.E., Gil, P.I., Martínez, L.C., Giordano, M.O. et al. 2015. Rotavirus seasonality in urban sewage from Argentina: effect of meteorological variables on the viral load and the genetic diversity. Environmental Research 138: 409–415.

Basnet, S., Palmenberg, A.C. and Gern, J.E. 2019. Rhinoviruses and their receptors. Chest 155(5): 1018–1025.

Battistone, A., Buttinelli, G., Bonomo, P., Fiore, S., Amato, C., Mercurio, P. et al. 2013. Detection of enteroviruses in influent and effluent flow samples from wastewater treatment plants in Italy. Food and Environmental Virology 6(1): 13–22.

Begier, E.M., Oberste, M.S., Landry, M.L., Brennan, T., Mlynarski, D., Mshar, P.A. et al. 2008. An outbreak of concurrent echovirus 30 and coxsackievirus A1 infections associated with sea swimming among a group of travelers to Mexico. Clinical Infectious Diseases 47(5): 616–623.

Ben-Haim, S., Gacinovic, S. and Israel, O. 2009. Cardiovascular infection and inflammation. pp. 103–114. *In*: Seminars in Nuclear Medicine (Vol. 39, No. 2). WB Saunders.

Bergelson, J.M., St John, N., Kawaguchi, S., Chan, M., Stubdal, H., Modlin, J. et al. 1993a. Infection by echoviruses 1 and 8 depends on the alpha 2 subunit of human VLA-2. Journal of Virology 67(11): 6847–6852.

Bergelson, J.M., Chan, M., Solomon, K.R., St John, N.F., Lin, H. and Finberg, R.W. 1994b. Decay-accelerating factor (CD55), a glycosylphosphatidylinositol-anchored complement regulatory protein, is a receptor for several echoviruses. Proceedings of the National Academy of Sciences 91(13): 6245–6248.

Bessaud, M., Blondel, B., Joffret, M. L., Mac Kain, A. and Delpeyroux, F. 2018. Nomenclature and classification of the enteroviruses: story of a long history. Virologie 22(6): 289–303.

Bian, L., Wang, Y., Yao, X., Mao, Q., Xu, M. and Liang, Z. 2015. Coxsackievirus A6: a new emerging pathogen causing hand, foot and mouth disease outbreaks worldwide. Expert Review of Anti-infective Therapy 13(9): 1061–1071.

Bibby, K. and Peccia, J. 2013. Identification of viral pathogen diversity in sewage sludge by metagenome analysis. Environmental Science and Technology 47(4): 1945–1951.

Blaas, D. and Fuchs, R. 2016. Mechanism of human rhinovirus infections. Molecular and Cellular Pediatrics 3(1): 1–4.

Boehm, A.B., Fuhrman, J.A., Mrše, R.D. and Grant, S.B. 2003. Tiered approach for identification of a human fecal pollution source at a recreational beach: case study at Avalon Bay, Catalina Island, California. Environmental Science and Technology 37(4): 673–680.

Borchardt, M.A., Bertz, P.D., Spencer, S.K. and Battigelli, D.A. 2003. Incidence of enteric viruses in groundwater from household wells in Wisconsin. Applied and Environmental Microbiology 69(2): 1172–1180.

Bouseettine, R., Hassou, N., Bessi, H. and Ennaji, M.M. 2020. Waterborne transmission of enteric viruses and their impact on public health. pp. 907–932. *In*: Emerging and Reemerging Viral Pathogens. Academic Press.

Cadamuro, R.D., Viancelli, A., Michelon, W., Fonseca, T.G., Mass, A.P., Krohn, D.M.A. et al. 2021. Enteric viruses in lentic and lotic freshwater habitats from Brazil's Midwest and South regions in the Guarani Aquifer area. Environmental Science and Pollution Research 28(24): 31653–31658.

Calgua, B., Mengewein, A., Grunert, A., Bofill-Mas, S., Clemente-Casares, P., Hundesa, A. et al. 2008. Development and application of a one-step low cost procedure to concentrate viruses from seawater samples. Journal of Virological Methods 153(2): 79–83.

Calgua, B., Fumian, T., Rusinol, M., Rodriguez-Manzano, J., Mbayed, V.A., Bofill-Mas, S. et al. 2013. Detection and quantification of classic and emerging viruses by skimmed-milk flocculation and PCR in river water from two geographical areas. Water Research 47(8): 2797–2810.

Carstens, E.B. and Ball, L.A. 2009. Ratification vote on taxonomic proposals to the International Committee on Taxonomy of Viruses 2008. Archives of Virology 154(7): 1181–1188.

Casas, I., Palacios, G.F., Trallero, G., Cisterna, D., Freire, M.C. and Tenorio, A. 2001. Molecular characterization of human enteroviruses in clinical samples: comparison between VP2, VP1, and RNA polymerase regions using RT nested PCR assays and direct sequencing of products. Journal of Medical Virology 65(1): 138–148.

Cashdollar, J.L. and Wymer, L. 2013. Methods for primary concentration of viruses from water samples: a review and meta-analysis of recent studies. Journal of Applied Microbiology 115(1): 1–11.

Cesari, C., Colucci, M.E., Veronesi, L., Giordano, R., Paganuzzi, F., Affanni, P. and Tanzi, M.L. 2010. Detection of enteroviruses from urban sewage in Parma. Acta Biomed. 81(1): 40–6.

Choudhary, M.C. 2019. Echovirus Infection: Background, Pathophysiology, Epidemiology. EMedicine.

Connell, C., Tong, H.I., Wang, Z., Allmann, E. and Lu, Y. 2012. New approaches for enhanced detection of enteroviruses from Hawaiian environmental waters. PloS One 7(5): e32442.

Corpuz, M.V.A., Buonerba, A., Vigliotta, G., Zarra, T., Ballesteros Jr, F., Campiglia, P. and Naddeo, V. 2020. Viruses in wastewater: occurrence, abundance and detection methods. Science of the Total Environment 745: 140910.

Costan-Longares, A., Moce-Llivina, L., Avellón, A., Jofre, J. and Lucena, F. 2008. Occurrence and distribution of culturable enteroviruses in wastewater and surface waters of north-eastern Spain. Journal of Applied Microbiology 105(6): 1945–1955.

Dalldorf, G. and Sickles, G.M. 1948. An unidentified, filtrable agent isolated from the feces of children with paralysis. Science 108(2794): 61–62.

De Giglio, O., Caggiano, G., Bagordo, F., Barbuti, G., Brigida, S., Lugoli, F. and Montagna, M.T. 2017. Enteric viruses and fecal bacteria indicators to assess groundwater quality and suitability for irrigation. International Journal of Environmental Research and Public Health 14(6): 558.

Di Prinzio, A., Bastard, D.P., Torre, A.C. and Mazzuoccolo, L.D. 2022. Hand, foot, and mouth disease in adults caused by Coxsackievirus B1–B6. Anais Brasileiros de Dermatologia.

Ehlers, M.M., Grabow, W.O.K. and Pavlov, D.N. 2005. Detection of enteroviruses in untreated and treated drinking water supplies in South Africa. Water Research 39(11): 2253–2258.

Elrick, M.J., Pekosz, A. and Duggal, P. 2021. Enterovirus D68 molecular and cellular biology and pathogenesis. Journal of Biological Chemistry, 296.

Farkas, K., Hillary, L.S., Thorpe, J., Walker, D.I., Lowther, J.A., McDonald, J.E. et al. 2021. Concentration and quantification of SARS-CoV-2 RNA in wastewater using polyethylene glycol-based concentration and qRT-PCR. Methods and Protocols 4(1): 17.

Faustini, A., Fano, V., Muscillo, M., Zaniratti, S., La Rosa, G., Tribuzi, L. et al. 2006. An outbreak of aseptic meningitis due to echovirus 30 associated with attending school and swimming in pools. International Journal of Infectious Diseases 10(4): 291–297.

Fong, T.T. and Lipp, E.K. 2005. Enteric viruses of humans and animals in aquatic environments: health risks, detection, and potential water quality assessment tools. Microbiology and Molecular Biology Reviews 69(2): 357–371.

Fongaro, G., Padilha, J., Schissi, C.D., Nascimento, M.A., Bampi, G.B., Viancelli, A. et al. 2015. Human and animal enteric virus in groundwater from deep wells, and recreational and network water. Environmental Science and Pollution Research 22(24): 20060–20066.

Fongaro, G., Stoco, P.H., Souza, D.S.M., Grisard, E.C., Magri, M.E., Rogovski et al. 2021. The presence of SARS-CoV-2 RNA in human sewage in Santa Catarina, Brazil, November 2019. Science of The Total Environment, 778, 146198.

Gaaloul, I., Riabi, S., Harrath, R., Hunter, T., Hamda, K. B., Ghzala, A. B. et al. 2014. Coxsackievirus B detection in cases of myocarditis, myopericarditis, pericarditis and dilated cardiomyopathy in hospitalized patients. Molecular Medicine Reports 10(6): 2811–2818.

Gawlik, B., Tavazzi, S., Mariani, G., Skejo, H., Sponar, M., Higgins, T. et al. 2021. SARS-CoV-2 Surveillance Employing Sewage: Towards a Sentinel System. Technical Report, European Union.

Gerba, C.P. and Schaiberger, G.E. 1975. Effect of particulates on virus survival in seawater. Journal (Water Pollution Control Federation) 93–103.

Gholipour, S., Ghalhari, M.R., Nikaeen, M., Rabbani, D., Pakzad, P. and Miranzadeh, M.B. 2022. Occurrence of viruses in sewage sludge: A systematic review. Science of the Total Environment 153886.

Gifford, R. and Dalldorf, G. 1951. The morbid anatomy of experimental Coxsackie virus infection. The American Journal of Pathology 27(6): 1047.

Goyle, K.K. and Walling, A. 2002. Diagnosing pericarditis. American Family Physician 66(9): 1695.

Gregory, J.B., Litaker, R.W. and Noble, R.T. 2006. Rapid one-step quantitative reverse transcriptase PCR assay with competitive internal positive control for detection of enteroviruses in environmental samples. Applied and Environmental Microbiology 72(6): 3960–3967.

Haaheim, Lars R., John R. Pattison and Richard J. Whitley (eds.). 2002. A Practical Guide to Clinical Virology. John Wiley & Sons.

Haramoto, E., Katayama, H., Utagawa, E. and Ohgaki, S. 2009. Recovery of human norovirus from water by virus concentration methods. Journal of Virological Methods 160(1-2): 206–209.

Haramoto, E., Kitajima, M., Hata, A., Torrey, J.R., Masago, Y., Sano, D. et al. 2018. A review on recent progress in the detection methods and prevalence of human enteric viruses in water. Water Research 135: 168–186.

Harvala, H., Broberg, E., Benschop, K., Berginc, N., Ladhani, S., Susi, P. et al. 2018. Recommendations for enterovirus diagnostics and characterization within and beyond Europe. Journal of Clinical Virology 101: 11–17.

Hata, A., Matsumori, K., Kitajima, M. and Katayama, H. 2015. Concentration of enteric viruses in large volumes of water using a cartridge-type mixed cellulose ester membrane. Food and Environmental Virology 7(1): 7–13.

Henry, R. 2012. Etymologia: Coxsackievirus. Emerging Infectious Diseases 18(11): 1871–1871.

Honig, E.I., Melnick, J.L., Isacson, P., Parr, R., Myers, I.L. and Walton, M. 1956. An endemiological study of enteric virus infections: Poliomyelitis, Coxsackie, and orphan (ECHO) viruses isolated from normal children in two socio-economic groups. The Journal of Experimental Medicine 103(2): 247–262.

Horstmann, D.M. 1958. The new ECHO viruses and their role in human disease. AMA Archives of Internal Medicine 102(1): 155–162.

Husby, A., Pasanen, A., Waage, J., Sevelsted, A., Hodemaekers, H., Janssen, R. et al. 2017. CDHR3 gene variation and childhood bronchiolitis. Journal of Allergy and Clinical Immunology 140(5): 1469–1471.

Hyypiä, T. and Harvala, H. 2015, January 1. Echoviruses☆. ScienceDirect; Elsevier.

ICTV. 2009. ICTV Master Species List 2008. Available in: https://talk.ictvonline.org/files/master-species-lists/m/msl/706. Accessed in: May 28, 2022.

ICTV. 2021. Virus Taxonomy: 2021 Release. Available in: https://talk.ictvonline.org/taxonomy/. Accessed in: May 28, 2022.

Ikner, L.A., Gerba, C.P. and Bright, K.R. 2012. Concentration and recovery of viruses from water: a comprehensive review. Food and Environmental Virology 4(2): 41–67.

Instituto Brasileiro de Geografia e Estatística (IBGE). 2019. Panorama. Ibge.gov.br.

Joffret, M.L., Polston, P.M., Razafindratsimandresy, R., Bessaud, M., Heraud, J.M. and Delpeyroux, F. 2018. Whole genome sequencing of enteroviruses species A to D by high-throughput sequencing: application for viral mixtures. Frontiers in Microbiology 2339.

Katayama, H., Shimasaki, A. and Ohgaki, S. 2002. Development of a virus concentration method and its application to detection of enterovirus and Norwalk virus from coastal seawater. Applied and Environmental Microbiology 68(3): 1033–1039.

Kennedy, J.L., Turner, R.B., Braciale, T., Heymann, P.W. and Borish, L. 2012. Pathogenesis of rhinovirus infection. Current Opinion in Virology 2(3): 287–293.

Khetsuriani, N., LaMonte-Fowlkes, A., Oberste, S. and Pallansch, M.A. 2006, September. Enterovirus Surveillance—United States, 1970–2005.

King, A.M., Lefkowitz, E., Adams, M.J. and Carstens, E.B. (eds.). 2011. Virus Taxonomy: Ninth Report of the International Committee on Taxonomy of Viruses (Vol. 9). Elsevier.

Laine, P., Savolainen, C., Blomqvist, S. and Hovi, T. 2005. Phylogenetic analysis of human rhinovirus capsid protein VP1 and 2A protease coding sequences confirms shared genus-like relationships with human enteroviruses. Journal of General Virology 86(3): 697–706.

Larivé, O., Brandani, J., Dubey, M. and Kohn, T. 2021. An integrated cell culture reverse transcriptase quantitative PCR (ICC-RTqPCR) method to simultaneously quantify the infectious concentrations of eight environmentally relevant enterovirus serotypes. Journal of Virological Methods 296: 114225.

Lau, S.K., Yip, C.C., Tsoi, H.W., Lee, R.A., So, L.Y., Lau, Y.L. and Yuen, K.Y. 2007. Clinical features and complete genome characterization of a distinct human rhinovirus (HRV) genetic cluster, probably representing a previously undetected HRV species, HRV-C, associated with acute respiratory illness in children. Journal of Clinical Microbiology 45(11): 3655–3664.

Li, L. and Delwart, E. 2011. From orphan virus to pathogen: the path to the clinical lab. Current Opinion in Virology 1(4): 282–288.

Majer, A., McGreevy, A. and Booth, T.F. 2020. Molecular pathogenicity of enteroviruses causing neurological disease. Frontiers in Microbiology 11: 540.

Majumdar, M., Sharif, S., Klapsa, D., Wilton, T., Alam, M.M., Fernandez-Garcia, M.D. and Martin, J. 2018, October. Environmental surveillance reveals complex enterovirus circulation patterns in human populations. *In*: Open Forum Infectious Diseases (Vol. 5, No. 10). US: Oxford University Press.

Mattingly, T.W. 1965. Changing concepts of myocardial diseases. Jama 191(1): 33–37.

Metcalf, T.G., Melnick, J.L. and Estes, M.K. 1995. Environmental virology: from detection of virus in sewage and water by isolation to identification by molecular biology—a trip of over 50 years. Annual Review of Microbiology 49(1): 461–487.

Ming, H.X., Zhu, L. and Zhang, Y. 2011. Rapid quantification of infectious enterovirus from surface water in Bohai Bay, China using an integrated cell culture-qPCR assay. Marine Pollution Bulletin 62(10): 2047–2054.

Ministério da Saúde. 2021. Portaria GM/MS N° 888, de 4 de maio de 2021 - DOU - Imprensa Nacional. Www.in.gov.br.

Minor, P.D. 2021. Polioviruses (Picornaviridae). Encyclopedia of Virology 2: 688–696.

Mohapatra, S., Menon, N.G., Mohapatra, G., Pisharody, L., Pattnaik, A., Menon, N.G. et al. 2021. The novel SARS-CoV-2 pandemic: Possible environmental transmission, detection, persistence and fate during wastewater and water treatment. Science of the Total Environment 765: 142746.

Monpoeho, S., Maul, A., Mignotte-Cadiergues, B., Schwartzbrod, L., Billaudel, S. and Ferre, V. 2001. Best viral elution method available for quantification of enteroviruses in sludge by both cell culture and reverse transcription-PCR. Applied and Environmental Microbiology 67(6): 2484–2488.

Morosky, S., Wells, A.I., Lemon, K., Evans, A.S., Schamus, S., Bakkenist, C.J. et al. 2019. The neonatal Fc receptor is a pan-echovirus receptor. Proceedings of the National Academy of Sciences 116(9): 3758–3763.

Muehlenbachs, A., Bhatnagar, J. and Zaki, S.R. 2014. Tissue tropism, pathology and pathogenesis of enterovirus infection. The Journal of Pathology 235(2): 217–228.

Nascimento, M.D.A.D. 2011. Estudo da presença de enterovírus, poliovírus e adenovírus humanos em amostras de águas de mananciais de Florianópolis/SC.

National Center for Immunization and Respiratory Diseases (NCIRD). 2021. CDC. Hand Foot and Mouth Disease.

National Research Council. 2002. Biosolids Applied to Land: Advancing Standards and Practices. National Academies Press.

Oberste, M.S., Maher, K., Michele, S.M., Belliot, G., Uddin, M. and Pallansch, M.A. 2005. Enteroviruses 76, 89, 90 and 91 represent a novel group within the species Human enterovirus A. Journal of General Virology 86(2): 445–451.

Okoh, A.I., Sibanda, T. and Gusha, S.S. 2010. Inadequately treated wastewater as a source of human enteric viruses in the environment. International Journal of Environmental Research and Public Health 7(6): 2620–2637.

Ooi, M.H., Wong, S.C., Lewthwaite, P., Cardosa, M.J. and Solomon, T. 2010. Clinical features, diagnosis, and management of enterovirus 71. The Lancet Neurology 9(11): 1097–1105.

Palacios, G. and Oberste, M.S. 2005. Enteroviruses as agents of emerging infectious diseases. Journal of Neurovirology 11(5): 424–433.

Palmenberg, A.C. and Gern, J.E. 2015. Classification and evolution of human rhinoviruses. pp. 1–10. *In*: Rhinoviruses. Humana Press, New York, NY.

Payne, S. 2017. Family Picornaviridae. Viruses, 107–114.

Pellegrinelli, L., Binda, S., Chiaramonte, I., Primache, V., Fiore, L., Battistone, A. et al. 2013. Detection and distribution of culturable Human Enteroviruses through environmental surveillance in Milan, Italy. Journal of Applied Microbiology 115(5): 1231–1239.

Pepper, I.L., Brooks, J.P. and Gerba, C.P. 2006. Pathogens in biosolids. Advances in Agronomy 90: 1–41.

Phyu, W.K., Ong, K.C., Kong, C.K., Alizan, A.K., Ramanujam, T.M. and Wong, K.T. 2017. Squamous epitheliotropism of Enterovirus A71 in human epidermis and oral mucosa. Scientific Reports 7(1): 1–10.

Pianetti, A., Baffone, W., Citterio, B., Casaroli, A., Bruscolini, F. and Salvaggio, L. 2000. Presence of enteroviruses and reoviruses in the waters of the Italian coast of the Adriatic Sea. Epidemiology and Infection 125(2): 455–462.

Pozzetto, B. and Gaudin, O.G. 1999. Coxsackieviruses (Picornaviridae).

Prado, T., Fumian, T.M., Mannarino, C.F., Maranhão, A.G., Siqueira, M.M. and Miagostovich, M.P. 2020. Preliminary results of SARS-CoV-2 detection in sewerage system in Niterói municipality, Rio de Janeiro, Brazil. Memórias do Instituto Oswaldo Cruz, 115.

Presidência da República - Casa Civil - Subchefia para Assuntos Jurídicos. 2007. LEI No 11.445.

Price, W.H. 1956. The isolation of a new virus associated with respiratory clinical disease in humans. Proceedings of the National Academy of Sciences of the United States of America 42(12): 892.

Rajtar, B., Majek, M., Polanski, L. and Polz-Dacewicz, M. 2008. Enteroviruses in water environment—a potential threat to public health. Annals of Agricultural and Environmental Medicine 15(2).

Ramos-Alvarez, M. and Sabin, A.B. 1956. Intestinal viral flora of healthy children demonstrable by monkey kidney tissue culture. American Journal of Public Health and the Nations Health 46(3): 295–299.

Robinson, C.R., Doane, F.W. and Rhodes, A.J. 1958. Report of an outbreak of febrile illness with pharyngeal lesions and exanthem: Toronto, summer 1957—isolation of group A coxsackie virus. Canadian Medical Association Journal 79(8): 615.

Romanos, M.T.V., Santos, N.S.D.O. and Wigg, M.D. 2015. Virologia Humana (3rd ed.). Guanabara Koogan.

Romero, J. 1999. Reverse-transcription polymerase chain reaction detection of the enteroviruses: overview and clinical utility in pediatric enteroviral infections. Archives of Pathology and Laboratory Medicine 123(12): 1161–1169.

Romero, J.R. and Modlin, J.F. 28 Aug 2014. Coxsackieviruses, echoviruses, and newer enteroviruses. Bennett, J.E., Dolin, R. and Blaser, M.J. (eds.). Mandell, Douglas, and Bennett's Principles and Practice of Infectious Diseases. 8th ed. Philadelphia, PA: Elsevier Inc. Vol 2: 2080–90.

Romero, J.R. 2017. Enteroviruses. International Encyclopedia of Public Health 474–478.

Schlindwein, A.D., Rigotto, C., Simões, C.M.O. and Barardi, C.R.M. 2010. Detection of enteric viruses in sewage sludge and treated wastewater effluent. Water Science and Technology 61(2): 537–544.

Schuler, B.A., Schreiber, M.T., Li, L., Mokry, M., Kingdon, M.L., Raugi, D.N. et al. 2014. Major and minor group rhinoviruses elicit differential signaling and cytokine responses as a function of receptor-mediated signal transduction. PloS One 9(4): e93897.

Schvoerer, E., Ventura, M., Dubos, O., Cazaux, G., Serceau, R., Gournier, N. et al. 2001. Qualitative and quantitative molecular detection of enteroviruses in water from bathing areas and from a sewage treatment plant. Research in Microbiology 152(2): 179–186.

Schwab, K.J., De Leon, R. and Sobsey, M.D. 1996. Immunoaffinity concentration and purification of waterborne enteric viruses for detection by reverse transcriptase PCR. Applied and Environmental Microbiology 62(6): 2086–2094.

Schyns, J.F., Hoekstra, A.Y., Booij, M.J., Hogeboom, R.J. and Mekonnen, M.M. 2019. Limits to the world's green water resources for food, feed, fiber, timber, and bioenergy. Proceedings of the National Academy of Sciences 116(11): 4893–4898.

Simmonds, P., McIntyre, C., Savolainen-Kopra, C., Tapparel, C., Mackay, I.M. and Hovi, T. 2010. Proposals for the classification of human rhinovirus species C into genotypically assigned types. Journal of General Virology 91(10): 2409–2419.

Simmonds, P., Gorbalenya, A.E., Harvala, H., Hovi, T., Knowles, N.J., Lindberg, A.M. et al. 2020. Recommendations for the nomenclature of enteroviruses and rhinoviruses. Archives of Virology 165(3): 793–797.

Sinclair, W. and Omar, M. 2022. Enterovirus. PubMed; StatPearls Publishing.

Sooksawasdi Na Ayudhya, S., Laksono, B.M. and van Riel, D. 2021. The pathogenesis and virulence of enterovirus-D68 infection. Virulence 12(1): 2060–2072.

Steele, M. and Odumeru, J. 2004. Irrigation water as source of foodborne pathogens on fruit and vegetables. Journal of Food Protection 67(12): 2839–2849.

Tedcastle, A., Wilton, T., Pegg, E., Klapsa, D., Bujaki, E., Mate, R. et al. 2022. Detection of enterovirus D68 in wastewater samples from the UK between July and November 2021. Viruses 14(1): 143.

Templeton, M.R., Andrews, R.C. and Hofmann, R. 2008. Particle-associated viruses in water: impacts on disinfection processes. Critical Reviews in Environmental Science and Technology 38(3): 137–164.

Tracy, S., Chapman, N.M. and Mahy, B.W.J. 2013. The Coxsackie B Viruses. In Google Books. Springer Science and Business Media.

Tschöpe, C., Ammirati, E., Bozkurt, B., Caforio, A.L., Cooper, L.T., Felix, S.B. et al. 2021. Myocarditis and inflammatory cardiomyopathy: current evidence and future directions. Nature Reviews Cardiology 18(3): 169–193.

Tulchinsky, T.H. 2018. John Enders, Jonas Salk, Albert Sabin and Eradication of Poliomyelitis. Case Studies in Public Health, 383–406.

Turner, R.B. 2007. Rhinovirus: more than just a common cold virus. The Journal of Infectious Diseases 195(6): 765–766.

UNESCO. 2017. 2017 - Wastewater, The Untapped Resource | United Nations Educational, Scientific and Cultural Organization. Unesco.org.

United States Environmental Protection Agency. 2020. Biosolids Laws and Regulations. www.epa.gov.

Victoria, M., Guimarães, F., Fumian, T., Ferreira, F., Vieira, C., Leite, J.P. et al. 2009. Evaluation of an adsorption–elution method for detection of astrovirus and norovirus in environmental waters. Journal of Virological Methods 156(1-2): 73–76.

Wiley, C.A. 2020. Emergent viral infections of the CNS. Journal of Neuropathology and Experimental Neurology 79(8): 823–842.

World Health Organization. 2003. Guidelines for environmental surveillance of poliovirus circulation (No. WHO/VandB/03.03). World Health Organization.

World Health Organization (WHO). 2019. Water. United Nations.

Wurtzer, S., Prevost, B., Lucas, F.S. and Moulin, L. 2014. Detection of enterovirus in environmental waters: a new optimized method compared to commercial real-time RT-qPCR kits. Journal of Virological Methods 209: 47–54.

Yin, Z., Voice, T.C., Tarabara, V.V. and Xagoraraki, I. 2018. Sorption of human adenovirus to wastewater solids. Journal of Environmental Engineering 144(11): 06018008.

Yin-Murphy, M. and Almond, J.W. 1996. Picornaviruses. Medical Microbiology. 4th edition.

Zell, R., Delwart, E. and Gorbalenya, A.E. 2017. ICTV virus taxonomy profile: Picornaviridae. Journal of General Virology 98(10): 2421–2422.

Chapter 3

Environmental Viruses in Livestock Production

Janice Reis Ciacci Zanella

‖‖‖

1. Introduction

Human beings have always depended on animals for food, transportation, work, and companionship. However, these animals can be a source of infectious diseases caused by viruses, bacteria, fungi and parasites, some of which can be transmitted to the human population (Seimenis 2008, Clemmons et al. 2021). These diseases are called zoonosis (Brown 2003).

This chapter will address the risk factors for the emergence of zoonosis, especially viral in production animals. These factors are related to the host (humans, domestic or wild animals, vectors), the infectious agent (viruses, bacteria, fungi, parasites) and the environment. Studies have pointed out that regions with high biodiversity, such as tropical regions, are hotspots for the emergence of diseases and most of the spillover factors are due to human behavior. In addition, this chapter will provide possible measures for control and management of emerging or transboundary animal diseases (TAD) and future research needs for developing better control approaches.

2. Impact of emerging zoonosis and transboundary diseases

Agriculture and livestock are essential sectors for many economies and many times family farming is the sole source of protein for people (FAO 2022). Considering the factors of production, diseases are the greatest threats to the stability of production systems, as their impact exceeds 20% of the losses in animal production worldwide. The socioeconomic impacts caused by animal diseases lead to increase in poverty, since, today, one billion farmers survive on production (Vallat and Wilson 2003). The Food and Agriculture Organization of the United Nations - FAO estimates that

---

Embrapa Suínos e Aves, BR 153, Km 110, Distrito de Tamanduá, 89.715-899, Concórdia, SC, Brazil.
* Corresponding author: janice.zanella@embrapa.br

zoonosis contributes significantly to losses of over 30 million tons of milk annually, and this contributes to malnutrition and decreased resistance to disease in children and the elderly (Seimenis 2008). In addition to productivity declines, countries lose trade opportunities due to health status and do not receive investment (Vallat and Wilson 2003).

A report published by the United States Agency for International Development (USAID) indicated that more than 75% of emerging human diseases in the last century are of animal origin (USAID 2009). Among infectious diseases, viral diseases are most disturbing and difficult to control as they usually spread very fast and unlike other infectious diseases have no cost-effective or safe antiviral antibiotics/drugs for the treatment (Yadav et al. 2020).

The World Bank estimates that zoonotic diseases affect more than two billion people worldwide, causing more than two million deaths every year, resulting in outbreaks with significant impacts on public health and the economy (World Bank 2021). Between 1997 and 2009, six major outbreaks of fatal zoonotic diseases emerged. Diseases included Nipah Virus in Malaysia, West Nile Fever in the USA, SARS (Severe Acute Respiratory Syndrome) in Asia and Canada, Highly Pathogenic Avian Influenza (HPAI) Virus in Asia and Europe, BSE (Bovine Spongiform Encephalopathy) or Mad Cow in the US and UK, Rift Valley fever in Africa. All mentioned diseases originated and/or involved wildlife, which have cost the global economy at least $80 billion dollars. If these outbreaks had been avoided, the losses of $7 billion a year would not have occurred. In this scenario, the investments required for a One Health system were estimated to be between $2–4 billion per year, which was substantially below the average of $7 billion in losses due to the six major outbreaks of zoonotic diseases between 1997–2009, considering none of the outbreaks of the six diseases evolved into a pandemic. Thus, it has further been estimated that an annual investment of approximately $2–4 billion is required to build and operate systems for effective disease prevention and control in low- and middle-income countries. Success in preventing the onset of pandemics comes with an expected annual rate of return of 86% (World Bank 2021).

The impact of COVID-19 or Coronavirus Disease 2019 caused by the new coronavirus (SARS-CoV-2) pandemics on human health and the world economy was immense, adding up to over 30 trillion dollars (Jackson 2021). Nevertheless, the impact of COVID-19 on the food supply chain, and specifically the meat chain, was unexpected. The high incidence of COVID-19 in workers in meat processing plants has rapidly evolved to affect human, animal and environmental well-being in several countries. This has led to the closure of processing plants due to outbreaks, especially the pork and poultry industries. The reduction in slaughter resulted in agglomeration of animals on farms, which led to the reduction of food for the animals, their euthanasia and inappropriate disposal of carcasses. This has had an impact on animal welfare, producer income, supply and biosecurity (Marchant-Forde and Boyle 2020).

Recently an expert panel has been assembled to study the emergence of diseases, especially on the concept of One Health (OHHLEP 2022). This has been an initiative of quadripartite organizations such as FAO, OIE, WHO and UNEP.

OHHLEP One Health High-Level Expert Panel has a specific work plan which includes a One Health definition:

One Health is an integrated, unifying approach that aims to sustainably balance and optimize the health of people, animals and ecosystems. It recognizes the health of humans, domestic and wild animals, plants, and the wider environment (including ecosystems) are closely linked and interdependent. The approach mobilizes multiple sectors, disciplines and communities at varying levels of society to work together to foster well-being and tackle threats to health and ecosystems, while addressing the collective need for clean water, energy and air, safe and nutritious food, taking action on climate change, and contributing to sustainable development. The study of one health is complex and multidisciplinary, nonetheless offers solutions for mitigation and prevention of disease emergence, leading to cooperative measures and global responses (OHHLEP 2022).

3. One world, one health, one medicine and one welfare

What are the strategies to combat the transmission of these pathogens, what is the role of public health actors? Could this be a consequence of globalization? What are the effects of zoonotic pandemics on animal production? (Zanella 2016)

Probably the infectious agents that cause emerging and TADs already existed in the environment and had the chance to cause diseases under certain altered conditions. The transmission of the infectious pathogen can occur between wildlife, human and domestic animals, between animals and humans, or between wildlife, domestic animal(s) and humans. Though, the key source for maintenance and transmission of the infectious agents in the environment is determined by the zoonotic pool and spill-over and spill-back mechanisms (Yadav et al. 2020).

The lesson learned from the emergence of COVID-19, SARS and avian influenza, was that an infectious disease in one country is a threat to others as well (Bekedam 2006, Mishra et al. 2021). Global threats need a global response and new or re-emerging diseases must be contained when they arise. Early detection, reporting and sharing of information and pathogens with countries and the international community is key to prompt a response at national and global levels. The contingency of emerging diseases requires government commitment and international collaboration. Not to mention at the local level, where health, agriculture and environmental authorities must collaborate in a transparent manner. Losing markets for animal products is a reality when public health is at stake. Governments need to take this responsibility in prevention and preparedness, in surveillance and response, in biosecurity and infection control at the hospital level and ultimately in the treatment of infectious diseases.

In the last century, many infectious or parasitic diseases have emerged or re-emerged. These include COVID-19, Ebola, Dengue, Chikungunya, Zika, yellow fever, tuberculosis, SARS, MERS-CoV (Middle East Respiratory Syndrome – MERS Coronavirus), measles, smallpox, HIV-AIDS, flu (human, avian or swine influenza), parasitic infections (trypanosomiasis), with more than 75% of them originating from animal microbial agents and can be a threat as biological weapons in bioterrorism (Tumpey et al. 2002, Seleem et al. 2009). One of the most likely factors to explain the recent occurrence of new diseases is the expansion of the human population (Panda et al. 2008, Mishra et al. 2021). Despite the concern about the scarcity of natural resources and the environment, it is estimated that we will be 10 billion people in the year 2050 (Brown 2003). This estimate is accompanied by a shocking increase in population urbanization from 39% in 1980 to 46% in 1997 and predicted to be 60% in 2030, which means an increase in human density in urban centers (Cutler et al. 2010).

4. Factors for the emergence or reemergence of zoonosis

In addition to the increase in human population, other global factors such as immense trade and travel, changes in the Earth's habitat, pollution and the expansion of animal production have favored the emergence of zoonotic disease agents (Zanella 2016). The favorable risk factors for these events to occur include all those mentioned in Figure 1 and are described below.

1) *Animal production and change in management practices*: Many hypotheses point to the expansion of livestock as a source of pathogenic agents for humans (Graczyk et al. 2000, Panda et al. 2008, LeJeune and Kersting 2010). There is a growing demand worldwide for animal protein and an increase in meat-producing animals in confinement (Guo et al. 2015). However, excessive

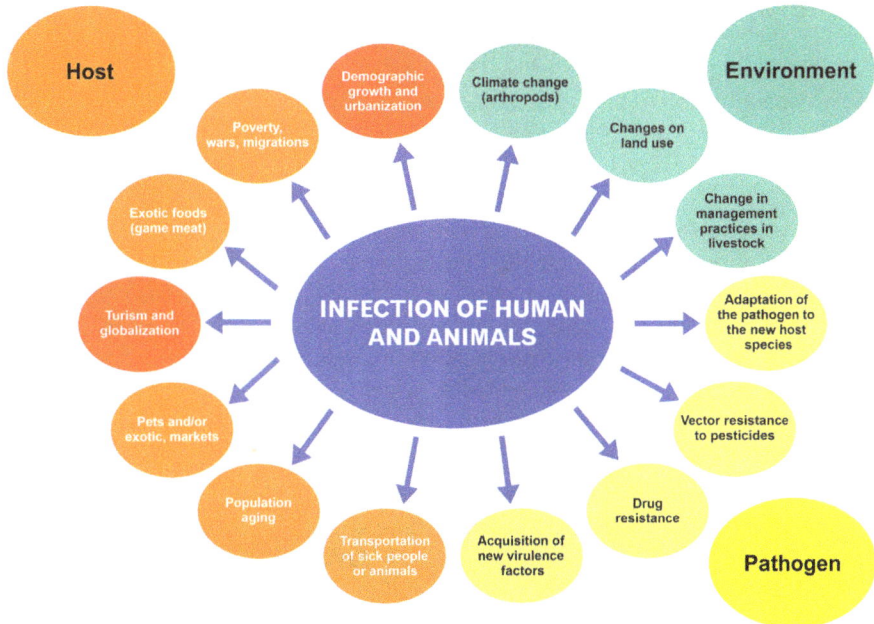

Figure 1. Factors influencing new and reemerging zoonosis (adapted from Cutler et al. 2010).

feedlots and nutrient processing for cattle feed may have led to the emergence of bovine spongiform encephalitis (Jacobson et al. 2009) or mad cow disease in the UK (Cutler et al. 2010). The mixing of animals of different species and under stressful conditions also favored the emergence of SARS in Asia (Cutler et al. 2010).

2) *Domestication and interaction with wild animals*: Agriculture may have changed the ecology of transmission of pre-existing human pathogens, resulting in new interactions between humans and wildlife. But, the domestication of animals has provided a conducive environment for infection of wild animal diseases in humans (Pearce-Duvet 2006). A phylogenetic study of zoonotic agents showed that both transmission from animal origin (domestic or wild) to humans and human to animals occurs. In particular, in the case of tuberculosis and taeniasis, the current evidence makes it difficult to determine the domestic origin, on the contrary, the origin and transmission is from human to animals (Pearce-Duvet 2006, Chomel 2007). Domestication of wild animals has led to the reemergence of zoonosis such as bovine tuberculosis in captured deer populations (Chomel 2007). Deer living in low population density and in the wild are less susceptible to disease. However, diseases become a threat when they are captured (or managed intensively), as wild animals can become reservoirs and recontaminate domestic animals. Wild swine such as wild boar, considered reservoirs of *Brucella suis biovar 2* infected feral (free-living) swine (Chomel 2007). Bats serve as a reservoir to numerous infectious agents, especially viruses. They are diverse, abundant mammalian species and can fly. In bats, more than 130 types of viruses have been recognized, including around 60 types of zoonotic viruses which are extremely pathogenic in humans, for example, rabies, SARS-like coronavirus, Ebola, Nipah, and Hendra viruses (Gupta et al. 2021).

3) *Acquisition of new virulence factors*: A new pathogenicity includes an increase in the potential for invasion, diffusion, toxin production or resistance to antimicrobial drugs. Pathogenic agents, including viruses, can undergo mutations or modifications in order to adapt to the human host. New viruses (emerging or re-emerging) are capable of rapid transmission because there is no host immune response or vaccines available. The emergence of viruses in fruit bats in the region of Australia and Southeast Asia caused diseases in animals and humans such as Hendra, Menangle, Lyssavirus and Nipah viruses. The latter emerged in Malaysia in 1999, decimating the swine industry and causing hundreds of deaths. Although little pathogenic for swine, Nipah virus causes severe disease in humans with respiratory and nervous signs, killing 40% of infected people (Brown 2003).

Infectious agents cross the species create barrier by using cell receptors, multiplying in the human host and causing the disease (Suarez Fernandez 2005, Van Poucke et al. 2010). These favorable conditions include the existence of vectors, infectious load, phylogenetic proximity between species and adaptability by mutation, recombination, as with influenza viruses, agents that cause influenza (*antigenic drift* or *shift,* for example) (Meulemans 1999, Alexander and Brown 2000). There is a threat that the highly pathogenic H5N1 avian influenza virus (H5N1-HP) will cause an epidemic if it crosses this barrier (Suarez Fernandez 2005, Van Reeth 2007), as there are permissible receptors in the human gut epithelium for infection by H5N1-HP, if meat from these infected animals is consumed (Arzt et al. 2010). World pandemics such as the Spanish flu of 1918 were caused by an H1N1 influenza virus of avian origin, where the H5N1-HP originating from a genomic rearrangement (*reassortment*) between two different virus strains (Harboe 1976, Hsieh et al. 2006, Vincent et al. 2008). The geographic distribution has increased of H5N1-HP has increased since 2003 because of hundreds of cases and deaths caused by the virus (Asia, Europe, Middle East, and West Africa) and the number of affected species has also increased (domestic and wild birds for cats, tigers and humans). As it is an unstable virus (constantly mutating), versatile and unpredictable, it is a great risk (Bekedam 2006). In Egypt alone, H5N1-HP caused 11 deaths in 2014, and in January 2015 alone, 12 new cases of H5N1-HP were recorded in humans in that country (Arafa et al. 2015, ProMED-mail 2015). The H7N9 avian influenza virus that emerged in China in March 2013 has already caused approximately 500 cases (as of January 2015) of respiratory problems, most of them severe, and transmission occurred through exposure to birds or contaminated environments (Lam et al. 2015). There is no evidence of human-to-human transmission of the new H7N9 in China, however it has pandemic potential due to the characteristic of constant evolution of influenza viruses (Yuan et al. 2015, Millman et al. 2015). The novel 2009 H1N1 (pH1N1) pandemic influenza virus originates from a reassortment between influenza viruses from three species: swine, human, and avian (Garten et al. 2009). The swine is a key species in the study of influenza virus rearrangements as it has receptors for both viruses of human, avian and swine origin in the cells of its respiratory system (Shinya and Kawaoka 2006). In the swine respiratory system, influenza viruses of different species (human or avian), in addition to swine, can infect, rearrange and replicate, giving rise to a new progeny of emerging viruses (Chen et al. 2011).

4) *Adaptation of the pathogen to the new host species*: Once-resistant populations are becoming susceptible to new, previously harmless pathogens, disrupting the balance of immune defense and infections (Slingenbergh 2021). This is due to the aging of the world population and the use of therapies that modulate the susceptibility to diseases due to immunodeficiency or immunosuppression, (Cutler et al. 2010).

5) *Changes in the environment.* Climate warming, the exploration of new agricultural frontiers, the introduction of vectors such as rodents and mosquitoes in urban areas change the dynamics of disease transmission. An example of this is the Ebola virus, identified in 1976 in central Africa, re-emerged in early 2014 in West African countries (Liberia, Sierra Leone, Ghana and Nigeria) with more than 20,000 cases since then. In the current epidemic, the index case resulted from exposure to a colony of insectivorous bats (*Mops condylurus*) (Mari Saez et al. 2014). The reduction in the abundance of natural hosts makes the vectors look for alternative hosts, increasing the opportunities for the transmission of diseases, such as the increase in human cases of Borreliosis or Lyme disease,

Ehrlichiosis and Anaplasmosis (Cutler et al. 2010). The development of areas close to tropical forests for the exploitation of rubber is related to the increase of schistosomiasis (Cutler et al. 2010). Another typical example of deforestation, invasion of wild habitat by urban development and mining activities are risk factors associated with the re-emergence of bats transmitting the rabies virus in humans in the Amazon Basin in 2004, where 46 people died (Chomel 2007). Climate change or human intervention in the environment has also been a cause of the emergence of diseases in animals such as the Rift Valley viral fever outbreak in Africa in 1987, 1997–1998 and 2006–2007, which were associated with changes in the course of the river and in floods as a result of dams or torrential rains in the region (Cutler et al. 2010).

6) *Transport of sick people and animals*: Infectious agents inside mosquitoes or animals can circle the planet in 24 hours and cause new outbreaks of disease. Whereas in the 19th century it took several weeks or months for disease agents to travel from one continent to another, today they can be transported to more distant lands in less time than the incubation period of many diseases (Brown 2003). Furthermore, the globalization of technology, information and the economy is creating market forces for industries, cultures and organisms to connect. Geopolitical barriers no longer exist. An example of this is the appearance of West Nile Virus (*West Nile* Virus) in the Western Hemisphere causing outbreaks on the east coast of the United States. The globalization of trade may have been responsible for the entry of the vector of this virus, *Aedes albopictus,* into the US state of Texas through an import of used tires from Japan and is currently widespread in that country (Brown 2003).

Traffic of people, live animals or animal products continues to grow (Seimenis 2008). Despite not being a classic zoonosis, foot-and-mouth disease was introduced in Brazil in 1870 with the importation of cattle from Europe, where the disease had been known since 1546 (Buainain 2007). As with foot-and-mouth disease, almost all animal diseases introduced in Brazil were of European origin, and from 1960 to 1980 more than 50 "new" diseases were identified in the country (MAPA 2011). The origin and provenance of animal protein can be studied through research on the stable isotopes of carbon and nitrogen in meat, and it also gives us the possibility to establish how meat produced in different countries travels around the world (Martinelli et al. 2011). Another way of transporting animals includes using them for sports activities such as hunting, fishing, horse racing, among others (Cutler et al. 2010).

7) *Tourism and Globalization*: Currently, approximately 2.5 million people use airports per day, with ≥ 1 million of these trips to international destinations, aggravating the transmission speed of agents (Cutler et al. 2010). World trade has tripled in the last 20 years and tourism is a champion growth sector in the global economy, where one in four citizens of a developed country visit a foreign country each year (Brown 2003). Ecotourism is the fastest growing tourism segment (average 10% per year) and includes safaris, extreme sports, tours, and exposure to habitats other than urban (Chomel 2007). Zoonosis associated with these practices include a variety of rickets, brucellosis, hepatitis E, hantaviruses,

leptospirosis, tick-borne encephalitis and schistosomiasis (Cutler et al. 2010). A recent example was the emergence of MERS-CoV caused by a coronavirus in Saudi Arabia in June 2012 (CDC 2015, Feikin et al. 2015). As of January 2015, 835 cases had been reported, with 358 deaths (43% mortality). MERS-CoV has been detected in bats and camels (Gossner et al. 2014, Sabir et al. 2016). The CDC (United States Center for Disease Control and Prevention) recommendation for travelers is to avoid contact with camels and the consumption of raw milk and undercooked meat from these animals (Gautret et al. 2013, Fanoy et al. 2014, CDC 2015, Sabir et al. 2016).

8) *Companion animals*: The presence of pets is very popular in homes around the world. It is estimated that in the United States more than half of households keep a pet cat or dog and 2.5% keep an exotic animal such as a ferret, rabbit, rodent or reptile (Chomel 2007). Keeping pets can be a risk for diseases, including scabies, salmonellosis, larva *migrans* and other parasites, viruses or bacteria present in the organisms of these animals and that can be transmitted to humans, mainly immunocompromised (Une and Mori 2007, Okulewicz and Bunkowska 2009). The bubonic plague is an example of a disease that has re-emerged in the western United States. The greatest risk of this plague comes from cats that get infected when hunting rodents as a result of exposure to fleas infected with *Yersinia pestis* (Cutler et al. 2010).

9) *Exotic pets, wet markets, contacts with zoo and circus animals*: Worldwide, it is estimated that 40,000 primates, 4 million birds, 640,000 reptiles and 350 million ornamental tropical fish are traded annually, generating an international wildlife trafficking industry estimated at 6 billion dollars (Chomel 2007). An example of the zoonosis transmitted by these activities include rabies, tuberculosis, brucellosis, psittacosis (*Chlamydia psittaci*) and H5N1-HP (Chomel 2007). Humans, especially children, can be exposed to zoonosis when participating in animal fairs (including agricultural fairs), zoos (especially mini-zoos) and circuses where infections by *Escherichia coli* O157:H7, *Salmonella, M. tuberculosis, Coxiella burnetii* and Influenza viruses have already been reported (Chomel 2007, Vincent et al. 2009). Exotic pets are also sources of illness for humans such as primate poxvirus infections in prairie dogs in the United States (at least 71 cases have been reported) (Chomel 2007). According to the CDC, 7% of human *Salmonella* infections in the US (mostly in children) are related to owning or having had contact with a pet reptile (Sanyal et al. 1997, Mermin et al. 2004). Studies suggest that the Wuhan market (Huanan Seafood Wholesale Market) was the epicenter of the start of COVID-19 pandemic (Maxmen 2022). Although no study has proven which animal the coronavirus SARS-CoV-2 came from, a hypothesis is that it was raccoon dogs, mammals used for food and their fur in China. Studies indicate that people who were in the session of these raccon dogs became more infected (Maxmen 2022).

10) *Exotic foods (game meat)*: Game meat consumption in many parts of the world, but especially in Central Africa and the Amazon Basin where up to 3 million tonnes and up to 167 thousand tonnes, respectively, are consumed annually (Chomel 2007). The Simian Foamy Virus is a zoonotic retrovirus very close

to the human immunodeficiency virus (HIV) and has already infected people who have had direct contact with raw meat of non-human primates. Similarly, other retroviruses such as human T-lymphotropic viruses type 3 and 4 have also been found in people who hunt, or keep monkeys as pets in Southern Cameroon (Chomel 2007). Trichinellosis is associated with consumption of game meat, such as bear meat, and recently severe cases of hepatitis E are associated with consumption of meat from deer and wild swine (wild boar). Other examples include parasites such as protozoa (Toxoplasma), trematodes (*Fasciola* sp., *Paragonimus* spp.), cestodes (*Taenia* spp., *Diphyllobothrium* sp.), and nematodes (*Trichinella* spp., *Anisakis* sp., *Parastrongylus* spp.) (Chomel 2007). The hunting industry in Africa generates billions of dollars annually and thus increases the frequency of exposure to primate retroviruses and other disease-causing agents such as the Ebola virus. The H5N1-HP that started in Southeast Asia in 2003 is a typical example of an agent transmitted by these animal markets. Another example is the SARS virus, which has been linked to trade in wild carnivores such as the Civet cat in China and MERS-Cov in camel meat and milk.

11) *Industrialization and food safety*: Emerging infectious agents can also be transmitted through food of animal origin. Bovine spongiform encephalitis or mad cow disease emerged in cattle in the UK (Jacobson, Lee et al. 2009) and several cases of a variant of the disease in humans, called Creutzfeldt-Jakob disease (vCJD) caused by transmission of BSE to people (Momcilovic and Rasooly 2000). Thus, from an emerging animal disease it quickly became an emerging zoonotic disease with hundreds of cases in the UK and consequently elsewhere in the world.

5. Transboundary animal diseases

Transboundary animal diseases (TAD) pose a serious risk to world agriculture and food security and undermine international trade (Clemmons et al. 2021). The world has faced devastating economic losses from large outbreaks of TADs such as foot and mouth disease, classical swine fever, African swine fever (ASF). Among the TADs having zoonotic manifestations, a number of infectious diseases, such as highly pathogenic avian influenza (HPAI due to the H5N1 virus), BSE (Mad cow disease caused by prion), West Nile fever, Rift Valley fever, SARS coronavirus, Hendra virus, Nipah virus, Ebola virus, Zika virus and Crimean-Congo hemorrhagic fever (CCHF), Middle East respiratory syndrome coronavirus (MERS-CoV), to name a few, adversely affecting animal and human health have been in the news in recent times (Yadav et al. 2020). Other zoonotic pathogens include Hemorrhagic Septicemia by Pasteurella multocida and Newcastle disease virus (NDV) (Clemmons et al. 2021). These diseases threatened decades ago and continue to do so.

In swine, the emergence in 2009 of the pandemic virus H1N1 changed the behavior of respiratory diseases, as the swine influenza virus, previously in equilibrium in swine herds, acquired sequences of the pandemic virus and evolved more frequently, making its difficult to control. The emergence of swine epidemic diarrhea or PED (Porcine Epidemic Diarrhea) in North America has caused enormous damage to the

American swine industry. African swine fever has dispersed and threatened swine production in Europe. Recently, Senecavirus caused outbreaks in swine farms in China, Brazil and the United States, causing considerable economic losses because it is an emerging agent and causes of vesicular disease. With regard to PRRS (swine reproductive and respiratory syndrome) in which very few countries are free, there are many studies of transmission between herds, but there are few that evaluate these transboundary transmission routes. Countries that remain PRRS negative typically apply severe restrictions on the importation of live pigs and semen. The risks for other routes of introduction such as aerial or local propagation, contact with wild pigs such as wild boar, consumption of washing or garbage containing meat or pork derivatives are still controversial or unknown. It is known that the transmission of diseases between animal species and between wild and domestic animals is a threat to animal health.

Many of these agents can also be transmitted by fomites (food, objects, vehicles), ticks, such as ASF virus and are highly stable in the environment (Clemmons et al. 2021). The greatest risks of spreading ASFV are through meat or meat products, food scraps, fomites (vehicles, flies, rats, water, and boots) and contact between live domestic and/or wild animals. Another long-term viable agent in the environment is Lumpy Skin Disease (LSD) virus. LSD causes disease in cattle (Bos indicus and Bos taurus) and buffalo (Bubalus bubalis), with several wild species being susceptible or seropositive. Transmission occurs by vectors such as blood-sucking arthropods, such as stable flies (Stomoxys calcitrans), mosquitoes (Aedes aegypti), horseflies (Haematopota spp.). Most outbreaks occur in summer, when arthropods are most active. The disease is enzootic in most African countries and some Middle Eastern countries, among others. The disease threatens international trade and can be used as an agent of economic bioterrorism (Clemmons et al. 2021). Another example of a resistant and persistent agent in meat is the swine vesicular disease virus (SVDV). Pigs are the natural hosts of SVDV that spreads through contact with other pigs or body fluids, or with contaminated fomites. Infected pigs shed the SVDV one to two days after infection, with shedding continuing for weeks, and the shed virus remaining present in the surrounding environment. As the disease is usually mild, food products from infected animals are more likely to enter the food chain and disseminate (Clemmons et al. 2021).

Survival of important viral pathogens in livestock was evaluated using trans-Pacific or trans-Atlantic transboundary models involving representative food ingredients, transport times and environmental conditions. The simulation used eleven viruses (or similars) including foot-and-mouth disease virus (FMDV), classical swine fever virus (CSFV), African swine fever virus (ASFV), swine influenza A virus (IAV-S), Pseudorabies (PRV), Nipah Virus (NiV), Porcine Reproductive and Respiratory Syndrome Virus (PRRSV), Swine Vesicular Disease Virus (SVDV), Vesicular Stomatitis Virus (VSV), Porcine Circovirus Type 2 (PCV2) and Vesicular Rash Swine Virus (VESV). Notably, more viruses survived in conventional soybean meal, lysine hydrochloride, choline chloride, vitamin D and pork sausage casings posing a risk for the transport of pathogens on a domestic and global level (Dee et al. 2018).

6. Action strategy, safety measures and disease control

Predicting the emergence (or return) of epidemics is difficult (Zanella 2016). However, the key point in preventing emerging zoonosis is to carry out early identification of pathogens in animals and rapid response before the disease becomes a threat to the human population. In this scenario, tools such as surveillance and study of viral evolution, which consist of studying the diversity and identification of genes of organisms, in general microbial, directly in their respective environments through sequencing are strategic. Surveillance of zoonotic agents helps in the control and informs health authorities of the possibility of the emergence of pathogens with pandemic potential, avoiding losses in production and in human and animal health. It is important to implement surveillance also for drug-resistant bacteria and viruses. Use a risk analysis-based strategy and invest in animal health advocacy, training and response to focus on geographic areas where these threats are likely to emerge. Research in applied molecular epidemiology will have immense value in the future, recognizing the associations between host and pathogen genotypes. In addition, research capable of generating knowledge and tools, combined with partnerships, is a great instrument for solving health problems for animal production chains. Thus, knowledge, surveillance, communication and cooperation are the forces that must come together to propose strategies and prevent countries, mainly those located on known hotspots from being the cradle of an emerging disease, which can harm the economy and health. The key is to have primary detection with control at risk borders (hotspots), rapid detection and control (slaughter of animals). Economic problems could arise with an embargo on meat from countries at risk or with a positive diagnosis, such as the foot-and-mouth disease epidemic in the UK in 2001, which caused approximately £6 billion in damage to the country, including losses to agriculture and tourism (Thompson et al. 2002).

At the University of Edinburgh in Scotland, 1,415 known human pathogens, 616 known farm animal pathogens, and 374 known carnivore pathogens were cataloged (Cleaveland et al. 2001). Of these human pathogens, 61.6% were zoonotic in origin. Of the 616 livestock pathogens, 77.3% were considered "multi-host" (infect more than one species), while 90% of the 374 carnivore pathogens were considered to be multi-host.

The OIE was founded in 1924 and currently has 182 member countries. In 2021, there were 264 OIE Reference Laboratories covering 109 diseases or topics in 37 countries and 65 Collaborating Centers in 31 countries (OIE 2022). Safety and disease control measures must be taken, this is the role of the veterinarian and ensuring transparency and preventing diseases from being transmitted from animals to humans is the duty of humankind. The OIE does not act in isolation, there have been agreements with the WHO since 1960 and in 2006 the FAO became part of this group. In addition, it is working to support the control of zoonosis emergence together with FAO, WHO and more recently, UNEP (OHHLEP 2022).

Specialists in human and animal health must build an early detection network for the disease at the local, regional and national levels (Zanella 2016). This network needs to have diagnostic laboratories, rapid disease contingency response and risk reduction. Strategies should focus on the following actions:

1) Detection of pathogens in wildlife (wild animals) that can cause disease in humans.

2) Analysis and characterization of potential risks and transmission methods of specific diseases of animal origin, verify the trade and transit of animals and products.

3) Institutionalization of the **One Health** strategy in various sectors that work in public health, including research and development of new drugs to treat emerging diseases (for example: antimicrobial-resistant bacteria).

4) Outbreak response that is appropriate at the country level, considering the stockpile of antivirals and how they will be distributed or having a strategy for obtaining them, likewise having a strategy for producing vaccines against pandemic diseases.

5) Risk reduction to prevent, minimize or eliminate the potential for emergence and transmission of new diseases.

The network should work on 5 plans that focus on surveillance (*Predict*), outbreak response investigation and training (*Respond*), laboratory network (*Identify*), formulate strategies to contain disease threats by studying their behavior (*Prevent*) and prepare for disasters and pandemics according to each situation (*Prepare*) (Zanella 2016). Rapid response includes timely disease report and compensation (Yadav et al. 2020). It is essential to foster partnerships with organizations that have experience in wildlife (fauna) monitoring, epidemiology and field training, have excellent laboratory infrastructure, good communication and national planning. It must have the support of different public and private sectors, strong cooperation and dedicated funding.

Sanitary and phytosanitary procedures and biosecurity are important for all human activities, including livestock health and production. Biosafety, biosecurity in all levels should be implemented. In addition, implementation of protocols and action plans for disease control such as risk assessment; communication and management; quarantine of imported animals and conducting structured disease surveillance and serosurveillance (Hulme 2020, Yadav et al. 2020). Effective inactivation protocols of viruses or other agents by composting or other treatments can reduce the risks of transmission by manure or water (Jeong et al. 2020).

In conclusion, several factors are important to be considered in the transmission of diseases or even in the emergence (or re-emergence) of diseases. However, three main aspects must be considered: host, agent and environment. Several threats and challenges are imposed on the production of food of animal origin, among them the globalization, which facilitated and intensified the food (and feed) trade. This liberalization of world trade, while offering many benefits and opportunities, also poses new risks. International organizations such as OIE and FAO seek to harmonize methods of diagnosis, detection, control and communication of diseases in order to reduce losses. The veterinary service should include routine surveillance, field investigations, sample collection, epidemiological investigations, risk analysis and mapping. It is imperative that these approaches exist at all levels, regional, national and international, and that an organizational structure is in place to better prevent and control animal and human health risks and the economic impact of emerging and transboundary animal diseases.

References

Alexander, D.J. and Brown, I.H. 2000. Recent zoonoses caused by influenza A viruses. Rev. Sci. Tech. 19(1): 197–225.

Arafa, A.S., Naguib, M.M., Luttermann, C., Selim, A.A., Kilany, W.H., Hagag, N. et al. 2015. Emergence of a novel cluster of influenza A(H5N1) virus clade 2.2.1.2 with putative human health impact in Egypt, 2014/15. Euro Surveill 20(13): 2–8.

Arzt, J., White, W.R., Thomsen, B.V. and Brown, C.C. 2010. Agricultural diseases on the move early in the third millennium. Vet Pathol. 47(1): 15–27.

Bank, T.W. 2021. Safeguarding Animal, Human and Ecosystem Health: One Health at the World Bank. From https://www.worldbank.org/en/topic/agriculture/brief/safeguarding-animal-human-and-ecosystem-health-one-health-at-the-world-bank.

Bekedam, H. 2006. Emerging Infectious Diseases, Defending against Infectious Diseases & Biological Threats. APEC.

Brown, C. 2003. Virchow Revisited: Emerging Zoonoses. ASM News 69(10): 5.

Buainain, A.M. 2007. Cadeia Produtiva de Carne Bovina Volume 8, Bib. Orton IICA/CATIE.

CDC. 2015. Middle East Respiratory Syndrome (MERS). Retrieved 2015-01-17, 2015, from http://www.cdc.gov/coronavirus/mers/.

Chen, L.M., Rivailler, P., Hossain, J., Carney, P., Balish, A., Perry, I. et al. 2011. Receptor specificity of subtype H1 influenza A viruses isolated from swine and humans in the United States. Virology 412(2): 401–410.

Chomel, B.B., Belotto, A. and Meslin, F.-X. 2007. Wildlife, exotic pets, and emerging zoonoses. Emerging Infectious Diseases 13, No. 1, January 2007(1): 6.

Cleaveland, S., Laurenson, M.K. and Taylor, L.H. 2001. Diseases of humans and their domestic mammals: pathogen characteristics, host range and the risk of emergence. Phil. Trans. R. Soc. Lond. B 356: 10.

Clemmons, E.A., Alfson, K.J. and Dutton III, J.W. 2021. Transboundary animal diseases, an overview of 17 diseases with potential for global spread and serious consequences. Animals 11(7): 2039.

Cutler, S.J., Fooks, A.R. and van der Poel, W.H. 2010. Public health threat of new, reemerging, and neglected zoonoses in the industrialized world. Emerg. Infect. Dis. 16(1): 1–7.

Dee, S.A., Bauermann, F.V., Niederwerder, M.C., Singrey, A., Clement, T., de Lima, M. and Diel, D.G. 2018. Correction: Survival of viral pathogens in animal feed ingredients under transboundary shipping models. PLoS One 13(11): e0208130.

Fanoy, E.B., van der Sande, M.A., Kraaij-Dirkzwager, M., Dirksen, K., Jonges, M., van der Hoek, W. et al. 2014. Travel-related MERS-CoV cases: an assessment of exposures and risk factors in a group of Dutch travellers returning from the Kingdom of Saudi Arabia, May 2014. Emerg. Themes Epidemiol. 11: 16.

FAO. 2022. Food and Agriculture Organization of the United Nations 2022, from https://www.fao.org/home/en/.

Feikin, D.R., Alraddadi, B., Qutub, M., Shabouni, O., Curns, A., Oboho, I.K. et al. 2015. Association of higher MERS-CoV virus load with severe disease and death, Saudi Arabia, 2014. Emerg. Infect. Dis. 21(11): 2029–2035.

Garten, R.J., Davis, C.T., Russell, C.A., Shu, B., Lindstrom, S., Balish, A. et al. 2009. Antigenic and genetic characteristics of swine-origin 2009 A(H1N1) influenza viruses circulating in humans. Science 325(5937): 197–201.

Gautret, P., Benkouiten, S., Salaheddine, I., Parola, P. and Brouqui, P. 2013. Preventive measures against MERS-CoV for Hajj pilgrims. Lancet Infect. Dis. 13(10): 829–831.

Gossner, C., Danielson, N., Gervelmeyer, A., Berthe, F., Faye, B., Kaasik Aaslav, K. et al. 2014. Human-Dromedary Camel Interactions and the Risk of Acquiring Zoonotic Middle East Respiratory Syndrome Coronavirus Infection. Zoonoses Public Health.

Graczyk, T.K., Evans, B.M., Shiff, C.J., Karreman, H.J. and Patz, J.A. 2000. Environmental and geographical factors contributing to watershed contamination with *Cryptosporidium parvum* oocysts. Environ. Res. 82(3): 263–271.

Guo, X., Raphaely, T. and Marinova, D. 2015. China's growing meat demands: implications for sustainability. Impact of Meat Consumption on Health and Environmental Sustainability: 221.

Gupta, P., Singh, M.P., Goyal, K., Tripti, P., Ansari, M.I., Obli Rajendran, V. and Malik, Y.S. 2021. Bats and viruses: a death-defying friendship. Virus Disease 32(3): 467–479.

Harboe, A. 1976 [Spanish influenza in 1918–1919 and the swine influenza virus]. Tidsskr Nor Laegeforen 96(16): 914.

Hsieh, Y.C., Wu, T.Z., Liu, D.P., Shao, P.L., Chang, L.Y., Lu, C.Y. et al. 2006. Influenza pandemics: past, present and future. J. Formos Med. Assoc. 105(1): 1–6.

Hulme, P.E. 2020. One Biosecurity: A unified concept to integrate human, animal, plant, and environmental health. Emerging Topics in Life Sciences 4(5): 539–549.

Jackson, J.K. 2021. Global economic effects of COVID-19, Congressional Research Service.

Jacobson, K.H., Lee, S., McKenzie, D., Benson, C.H. and Pedersen, J.A. 2009. Transport of the pathogenic prion protein through landfill materials. Environmental Science & Technology 43(6): 2022–2028.

Jeong, K.H., Lee, D.J., Lee, D.H., Ravindran, B., Chang, S.W., Mupambwa, H.A. and Ahn, H.K. 2020. Composting reduces the vitality of H9N2 in poultry manure and EMCV in pig manure allowing for an environmentally friendly use of these animal wastes: a preliminary study. Microorganisms 8(6): 829.

Lam, T.T., Zhou, B., Wang, J., Chai, Y., Shen, Y., Chen, X. et al. 2015. Dissemination, divergence and establishment of H7N9 influenza viruses in China. Nature 522(7554): 102–105.

LeJeune, J. and Kersting, A. 2010. Zoonoses: an occupational hazard for livestock workers and a public health concern for rural communities. J. Agric Saf. Health 16(3): 161–179. doi: 10.13031/2013.32041.

MAPA. 2011. Ministério da Agricultura, Pecuária e Abastecimento. Retrieved 18/03/2011, 2011, from http://www.agricultura.gov.br/portal/page/portal/Internet-MAPA/pagina-inicial/animal.

Marchant-Forde, J.N. and Boyle, L.A.J.F.i.v.s. 2020. COVID-19 effects on livestock production: A One Welfare issue. 734.

Mari Saez, A., Weiss, S., Nowak, K., Lapeyre, V., Zimmermann, F., Dux, A. et al. 2014. Investigating the zoonotic origin of the West African Ebola epidemic. EMBO Mol. Med. 7(1): 17–23.

Martinelli, L.A., Nardoto, G.B., Chesson, L.A., Rinaldi, F.D., Ometto, J.P.H., Cerling, T.E. et al. 2011. Worldwide stable carbon and nitrogen isotopes of Big Mac® patties: An example of a truly "glocal" food. Food Chemistry 127(4): 1712–1718.

Maxmen, A. 2022. Wuhan market was epicentre of pandemic's start, studies suggest. Nature 603(7899): 15–16.

Mermin, J., Hutwagner, L., Vugia, D., Shallow, S., Daily, P., Bender, J. et al. 2004. Reptiles, amphibians, and human Salmonella infection: a population-based, case-control study. Clin. Infect. Dis. 38 Suppl 3: S253–261.

Meulemans, G. 1999. [Inter-species transmission of the influenza virus]. Bull. Mem. Acad. R Med. Belg 154(5-6): 263–270; discussion 270–262.

Millman, A.J., Havers, F., Iuliano, A.D., Davis, C.T., Sar, B., Sovann, L. et al. 2015. Detecting spread of Avian Influenza A(H7N9) Virus beyond China. Emerg. Infect. Dis. 21(5): 741–749.

Mishra, J., Mishra, P. and Arora, N.K. 2021. Linkages between environmental issues and zoonotic diseases: with reference to COVID-19 pandemic. Environmental Sustainability 4(3): 455–467.

Momcilovic, D. and Rasooly, A. 2000. Detection and analysis of animal materials in food and feed. J. Food Prot. 63(11): 1602–1609.

OHHLEP. 2022. One Health High-Level Expert Panel. from https://www.who.int/publications/m/item/one-health-high-level-expert-panel-annual-report-2021.

OIE. 2022. World Organization for Animal Health, 2022, from https://www.oie.int/en/home/.

Okulewicz, A. and Bunkowska, K. 2009. [Baylisascariasis—a new dangerous zoonosis]. Wiad Parazytol 55(4): 329–334.

Panda, A.K., Thakur, S.D. and Katoch, R.C. 2008. Rabies: control strategies for Himalayan states of the Indian subcontinent. J. Commun. Dis. 40(3): 169–175.

Pearce-Duvet, J.M.C. 2006. The origin of human pathogens: evaluating the role of agriculture and domestic animals in the evolution of human disease. Biol. Rev. 81: 14.

ProMED-mail, 2015, 2015-02-24. Avian Influenza, Human (52): Egypt (DAKAHLIA) H5N1 Fatality. 2015.

Sabir, J.S., Lam, T.T., Ahmed, M.M., Li, L., Shen, Y., S, E.M.A.-A., Qureshi, M.I., Abu-Zeid, M. et al. 2016. Co-circulation of three camel coronavirus species and recombination of MERS-CoVs in Saudi Arabia. Science 351(6268): 81–84.

Sanyal, D., Douglas, T. and Roberts, R. 1997. *Salmonella* infection acquired from reptilian pets. Arch. Dis. Child 77(4): 345–346.

Seimenis, A.M. 2008. The spread of zoonosis and other infectious diseases through the international trade of animals and animal products. Veterinaria Italiana 44(4): 9.

Seleem, M.N., Boyle, S.M. and Sriranganathan, N. 2009. Brucellosis: a re-emerging zoonosis. Vet. Microbiol. 140(3-4): 392–398.

Shinya, K. and Kawaoka, Y. 2006. [Influenza virus receptors in the human airway]. Uirusu 56(1): 85–89.

Slingenbergh, J. 2021. Outer to inner-body shifts in the virus-host relationship for the three main animal host domains of the world today: wildlife, humans, and livestock. Immunome Research 17(1): 1–9.

Suarez Fernandez, G. 2005. [Natural history of Avian Influenza or "chicken flu". Present and future health analysis]. An R Acad. Nac. Med. (Madr) 122(2): 215–228; discussion 228–232.

Thompson, D., Muriel, P., Russell, D., Osborne, P., Bromley, A., Rowland, M. et al. 2002. Economic costs of the foot and mouth disease outbreak in the United Kingdom in 2001. Rev. Sci. Tech. 21(3): 675–687.

Tumpey, T.M., Garcia-Sastre, A., Mikulasova, A., Taubenberger, J.K., Swayne, D.E., Palese, P. et al. 2002. Existing antivirals are effective against influenza viruses with genes from the 1918 pandemic virus. Proc. Natl. Acad. Sci. U S A 99(21): 13849–13854.

Une, Y. and Mori, T. 2007. Tuberculosis as a zoonosis from a veterinary perspective. Comp. Immunol. Microbiol. Infect. Dis. 30(5-6): 415–425.

USAID. 2009. USAID Launches Emerging Pandemic Threats Program, Retrieved 08/08/2010, 2010, from http://www.usaid.gov/press/releases/2009/pr091021_1.html.

Vallat, B. and Wilson, D. 2003. Obligations of the member states of the World Animal Health Organization regarding the organization of their veterinary services. Revue Scientifique Et Technique De L Office International Des Epizooties 22(2): 553–559.

Van Poucke, S.G., Nicholls, J.M., Nauwynck, H.J. and Van Reeth, K. 2010. Replication of avian, human and swine influenza viruses in porcine respiratory explants and association with sialic acid distribution. Virol. J. 7: 38.

Van Reeth, K. 2007. Avian and swine influenza viruses: our current understanding of the zoonotic risk. Vet. Res. 38(2): 243–260.

Vincent, A.L., Ma, W., Lager, K.M., Janke, B.H. and Richt, J.A. 2008. Swine influenza viruses a North American perspective. Adv. Virus Res. 72: 127–154.

Vincent, A.L., Swenson, S.L., Lager, K.M., Gauger, P.C., Loiacono, C. and Zhang, Y. 2009. Characterization of an influenza A virus isolated from pigs during an outbreak of respiratory disease in swine and people during a county fair in the United States. Vet. Microbiol. 137(1-2): 51–59.

Yadav, M.P., Singh, R.K. and Malik, Y.S. 2020. Emerging and transboundary animal viral diseases: Perspectives and preparedness. Emerging and Transboundary Animal Viruses, Springer: 1–25.

Yuan, J., Lau, E.H., Li, K., Leung, Y.H., Yang, Z., Xie, C. et al. 2015. Effect of live poultry market closure on avian influenza A(H7N9) virus activity in Guangzhou, China, 2014. Emerg. Infect. Dis. 21(10): 1784–1793.

Zanella, J.R.C. 2016. Emerging and reemerging zoonoses and their importance for animal health and production. Pesquisa Agropecuária Brasileira 51: 510–519.

Chapter 4

Bacteriophages Discovery and Environmental Application

Raphael da Silva,[1] *Mariana Elois,*[1] *Beatriz Pereira Savi,*[1]
Estevão Brasiliense de Souza,[1] *Rafael Dorighello Cadamuro,*[1]
Doris Sobral Marques Souza,[1] *David Rodríguez-Lázaro*[2] and
Gislaine Fongaro[1,*]

1. Introduction

1.1 Bacteriophage discovery and early research

It is well accepted in the scientific literature that bacteriophage's discovery is an achievement carried out independently, by two researchers, far apart in different countries. In 1915, Frederick William Twort, an English bacteriologist, was studying if it was possible to grow different viruses (what in that time they usually referred to as filterable agents) in different types of medium. Twort did not succeed in this attempt, as all viruses depend on cells to replicate—a fact that scientists clearly know today. In one of his attempts, when he inoculated a not filtered sample, used for smallpox vaccination, on a agar medium, Twort observed that a bacteria had grown on it, but not in a normal way. He said in his report that "*inoculated agar tubes often showed watery-looking areas, and in cultures that grew micrococci it was found that some of these colonies could not be subcultured, but if kept they became glassy and transparent*" (Duckworth 1976, Twort 1915).

Twort even noticed that after the filtration through a Chamberland candle of this "glassy material" it still had the ability to cause that effect. He then came to the conclusion that such a phenomenon was caused by a filterable and infectious agent that had the capacity to multiply itself and kill the bacteria. Due to financial

[1] Laboratory of Applied Virology, Federal University of Santa Catarina, Department of Microbiology, Immunology and Parasitology, Santa Catarina State, Brazil.
[2] Instituto Tecnológico Agrario de Castilla y León, 47071 Valladolid, Spain.
* Corresponding author: gislaine.fongaro@ufsc.br

problems Twort could not continue his research. In his words: "*I regret that financial considerations have prevented me from carrying these researches to a definite conclusion*" (Twort 1915).

Another prominent person in the bacteriophage history is Félix d'Herelle, a French-Canadian bacteriologist. In 1910, d'Herelle was studying a disease that affected locusts, a species that had invaded the region, in the state of Yucatan, México. He found that the illness was caused by a bacterium and could demonstrate that by infecting health insects. During his research, Félix noticed some regions with small circular clear spots in some of the isolated bacterial cultures grown on agar. After repeated attempts to visualize in microscope what was causing that anomaly, d'Herelle concluded that such an agent must be so small that it could pass even through a Chamberland filter without being held back. Despite the intriguing events that occasionally appeared in his bacterial cultures, and due to the inconsistency of the appearance of these spots (together with the fact that Félix was unable to reproduce the phenomenon), d'Herelle couldn't study it at the time (Duckworth 1976).

When investigating an epidemic of dysentery that afflicted the french cavalry squadron in 1915, Félix filtered samples of feces from diseased men and cultured it with dysentery bacilli, noticing once again the clear spots. He then made the decision to follow the course of a patient's illness in an attempt to see when the clear spots started to appear. On the first day d'Herelle isolated a Shiga dysentery bacillus from the patient's bloody stool, a culture grew normally when he added the filtered sample. On the fourth day he repeated the experiment and let the culture incubate overnight, the next day in the morning he had a surprise, the broth culture was clear and not turbid for the bacterial growth and the agar plates were also clear (d'Herelle 1917, Duckworth 1976). It was at that moment that Félix realized that this could only be caused by a virus that was killing the bacteria, he thought that the patient must also be better from his infection at that time, because the event that occurred in his experiments must also have occurred within the patient. Upon visiting the patient, d'Herelle describes that the sick man did indeed improve that very night (d'Herelle 1917, Duckworth 1976). d'Herelle used the term bacteriophage, in which the prefix comes from the word "bacteria" and the suffix comes from the Greek word phagein that can be translated as "to devour".

When penicillin was discovered, phage studies were set aside, as such chemicals had a very relevant antimicrobial capacity. However, in Eastern European countries and in Russia (at the time Soviet Union) the research on phage therapy continued (Clokie et al. 2011, Salmond and Fineran 2015).

1.2 Bacteriophage diversity

Bacteriophages, or phages, are viruses that infect bacteria. They are ubiquitous and are the most abundant and diverse entities known on the planet. It is estimated that there are around 10^{31} viral particles globally (Suttle 2005). They are obligate intracellular parasites of prokaryotes, consisting of nucleic acid surrounded by a protein coat (capsid), which may or may not be surrounded by a lipoprotein envelope (Seed et al. 2013).

Phages have a high diversity when it comes to their particular characteristics. In terms of the structural morphology it can vary greatly among their groups. They can be icosahedral, polyhedral, filamentous, spherical, rod or even pleomorphic (Ackermann and Prangishvili 2012, Dion et al. 2020). The genetic material constitution can also show a lot of possibilities. The larger group of identified phages, *Caudovirales*, is the order that includes different families of tailed viruses, and possesses a genome of double strand DNA (dsDNA). Single strand DNA (ssDNA), single strand RNA (ssRNA) and dsRNA viruses have already been identified in nature. In terms of sizes, phages are usually in the scale of hundreds of nanometers, as the *Escherichia virus T4*, a myovirus with about 200 nm long (Dion et al. 2020).

Over 6000 bacteriophages have already been studied in electron microscopy, of which more than 96% belongs to the order *Caudovirales*, the families *Siphoviridae*, *Myoviridae* and *Podoviridae*, are the most abundant phages described, respectively (Ackermann and Prangishvili 2012). Genomic diversity correlates with virus data analyzed by electron microscopy; more than half of the complete phage genomes present in the National Center for Biotechnology Information (NCBI) belong to the *Siphoviridae*. The families *Myoviridae* and *Podoviridae* represent a relevant portion of the other half of the complete phage genomes kept in the database. The super representation of tailed phages, in proportion to others groups, may be due to biases in isolation methods and due to some projects that focus on the isolation and genome sequencing of specific groups. For example, Science Education Alliance–Phage Hunters Advancing Genomics and Evolutionary Science (SEA-PHAGES) programme isolated and characterized more than 1500 phages from *Siphoviridae* infecting *Mycobacterium smegmatis* (Dion et al. 2020).

The genomic diversity of bacteriophages is remarkable, however, the number of exaggerated genomes in the NCBI still represents a small fraction of the real potential of this phage diversity in different environments. Since there is no genetic marker, it is universal for viruses and the predicted number of phages in the biosphere is huge (Dion et al. 2020). The use of metagenomics facilitated the knowledge of the diversity of viruses in different media, in this approach, the total viral component of a given environment is collected and sequenced. It is possible that genomic data are collected, without the cultivation of hosts and isolation of the phages, as well as information on phages that are not susceptible to propagation or that do not have hosts in culture (Suttle 2007, Clokie et al. 2011).

The taxonomic classification of viruses is carried out by the International Committee on Taxonomy of Viruses (ICTV), the website of the committee also provides information on basic characteristics of different groups of viruses, and the Bacterial Viruses Subcommittee is responsible for phage taxonomy. On the ICTV taxonomy website page it is possible to find more than 48 families of bacteriophages, distributed in 14 orders, being the *Caudovirales* with 14 families, the group that has the most families, followed by *Norzivirales* with 4 families and *Tubulavirales* and *Halopanivirales* with 3 families each, the rest of the orders have 2 or less families. Genetic material content, virion morphology and genomic features are characteristics often used for phage classification (Ackermann 2009, Ye et al. 2019). Table 1 provides some examples to illustrate phage diversity.

Table 1. Bacteriophage diversity (ICTV).

Family	Genus	Genome	Envelope	Morphology	Main host
Myoviridae	*T4-like virus*	dsDNA	Absent	Long, straight and contractile tail.	*Enterobacteriaceae*
Siphoviridae	*Lambdavirus*	dsDNA	Absent	Long and non-contractile tail.	*Escherichia coli*
Podoviridae	*Lederbergvirus*	dsDNA	Absent	Short non-contractile tail.	*E. coli, Salmonella* and *Shigella*
Inoviridae	*Fibrovirus*	ssDNA	Absent	Rod or filamentous	*Vibrio cholera*
Microviridae	*Phix174microvirus*	ssDNA	Absent	Icosahedral	*Enterobacteriaceae*
Cystoviridae	*Cystovirus*	dsRNA	Present	Spherical	*Pseudomonas* spp.
Leviviridae	*Allolevivirus*	ssRNA+	Absent	Spherical	*Enterobacteriaceae*

Estimates vary depending on the environment, several studies suggest that for each bacterial cell in seawater there are 10 times more phage particles. There is a correlation between system productivity and phage abundance. In this sense, viruses that infect pelagic host species are the most abundant, while ambients away from the shore and at greater depths, see diminished number of bacterial viruses (Clokie et al. 2011, Martinez-Hernandez et al. 2017, Suttle 2005).

2. Potential phage application

It is not within the scope of this text to exhaust the subject, this topic is only intended to present the reader with an overview of the possible environmental phage applications. There are several biotechnological approaches for environmental phage application, each approach has its principles and purposes. Bacteriophages may be applied as a direct decontamination agent for different targets, as a biofilm breaker, as a vector of antibiotic susceptibility genes and as a method for bacteria detection, among other usages (Altamirano and Barr 2019, Santos and Azeredo 2019).

Simply, phages can present two types of replicative cycles, lytic and lysogenic, being referred to as virulent and temperate phages, respectively. Lytic cycle is characterized by fast steps of replication and for killing the cell after the process. When it comes to phage therapy (the use of bacteriophages to treat pathogen infections), ideally, viruses that present the lytic cycle and just that, are the ones that are selected for such application (Altamirano and Barr 2019). Phages that go through the lysogenic cycle are characterized for a slow replication (thus capable of being stable for generations of hosts), integration of the genetic material to the host's, and more important, by the transmission of antibiotic resistance genes (ARGs) (Clokie et al. 2011, O'Sullivan et al. 2019). *Caudovirales* and *Microviridae* are the groups with most phages that present the lytic cycle and infect human pathogens (Lin et al. 2017).

Whether viruses are living entities or not is still up for debate, but it is clear that phages go through natural selection and evolve, and that is the primordial difference from antibiotics (Altaminano and Barr 2019, Principi et al. 2019).

When scientists look for phages in order to isolate them, they tend to look in their host's natural environment (Weber-Dąbrowska et al. 2016). To give the reader an example, you can think of *Salmonella enterica,* which is a bacteria that belongs to the family *Enterobacteriaceae,* this group is characterized by Gram negative bacillus, fermenting glucose bacteria that inhabit the gastrointestinal tract of mammals (Octavia and Lan 2014). In order to find a phage capable of infecting and killing that pathogen, the scientists study the bacteria's habitat and its route of infection, as well as its environmental route. Thus, some samples that the researcher can look for are, for example, sewage samples, patient feces samples, animal waste samples, among others (Pelzek 2008, Weber-Dąbrowska et al. 2016).

Other features that differentiate the performance of antibiotics from phages is the specificity of action, antimicrobial chemicals tend to have a broad spectrum of action that can affect negatively bacterias relevants to natural microbiota of human. On the other hand, this does not mean that phages are free from this effect, however caution is needed when selecting and characterizing them as their host range. Phages also seems to be less expensive and more secure than antibiotics (Principi et al. 2019).

Biofilms are communities of bacteria, extremely and intimately organized, and even functionally organized. These communities are attached to surfaces and are protected by polymeric matrices produced by themselves. Biofilm is the form of organization that bacteria are most commonly found in nature, clinical and industrial settings, even more common than the free form (Costerton et al. 1995). This organization can even protect the community against the action of different antimicrobial agents by acting as an ion exchanger, controlling what comes in and out (Davey and O'Toole 2000). Phages and their hosts co-evolved for millions of years, so it is to be expected that they have found ways to degrade the biofilms formed by the bacteria. And in fact it does, there are several ways in which bacteriophages can invade, infect and destroy such bacterial colony organizations. One of these ways is the replication in susceptible hosts and subsequent expansion of the viral titer, these viruses then become concentrated within the biofilm, thus increasing their ability to infect new bacteria, which would produce more matrix, in addition to eliminating the bacteria that could regenerate the biofilm (Harper et al. 2014).

Another way that phages found to eliminate biofilms is through the production of enzymes called depolymerases. These enzymes are capable of carrying out the degradation of Extracellular Polymeric Substance (EPS), an important component in the composition of the biofilm matrix. The phages can act degradating EPS not only using external factors but, they can also induce internal degradation by bacterias. When it comes to the production of EPS degrading enzymes, phages capable of integrating their genome with the host can transmit genes that produce these enzymes, which can be transcribed and translated by the bacteria themselves, a real Trojan horse (Harper et al. 2014).

Inactive cells present within the biofilm matrix that can provide structural reconstruction are called persistent cells. Lysogenic phages have the ability to cause latent infections in these cells, as these viruses are not capable of replicating and destroying these bacteria. However, they can remain latent in the bacteria until their reactivation. When this occurs, the viruses can enter a lytic cycle and kill the cells, preventing the recomposition of the biofilm (Pearl et al. 2008).

Bacteriophage cocktails, that is, a solution containing multiple species of phage may be a more effective alternative in combating biofilms, not only for increasing the host range, but also for preventing or delaying the appearance of mutations that can cause resistance to bacteriophages, thus hindering the treatment (Gu et al. 2012, Tanji et al. 2004). A study demonstrated that pretreatment of catheters with bacteriophage cocktail for *Pseudomonas aeruginosa* was able to inhibit the development of the biofilm, the use of multiple phage species has the ability to increase the host range, thus infecting more species and variants present in the biofilm (Fu et al. 2010). In another study, Alves et al. (2016), isolated and characterized six phages capable of infecting *P. aeruginosa*. These phages were used in the formulation of a cocktail to be tested against *P. aeruginosa* biofilms in flow and static conditions. The cocktail proved to be highly efficient in controlling biofilms in both conditions, a higher Multiplicity of Infection (MOI) value was more effective in decreasing biomass and preventing biofilm regrowth. In the same way that phage cocktails can be a helpful tool, the use of antibiotics combined with phage cocktails, can optimize the removal of biofilms (Yilmaz et al. 2013).

Bioengineering allows the expansion of phage application mechanisms in biotechnology. Some examples of how bacteriophages can be used as a vector of genetic or chemical elements are, delivery of antibiotic sensitivity genes, which aim to reverse antibiotic resistance (Edgar et al. 2012), delivery of genes that actually enhance the antibiotic lethality (Ronayne et al. 2016) or even the delivery of antibiotics directly to bacterial cells, which is able to increase the efficiency of the antimicrobial (Yacoby et al. 2007).

Bacteriophages can also be applied in the detection of bacteria, phage approaches have some advantages over traditional bacterial detection methods. Large-scale production of these viruses is cheap and easy to develop, in addition, phages are ubiquitous entities, so it is possible to design detection techniques for virtually any bacteria (Richter et al. 2018). Another advantage presented by phages is that many phages are highly stable over wide variations and temperature (Brigati and Petrenko 2005). Phage replication and subsequent release and spreading of the internal elements of the bacteria can be used as a signal trigger, a marker of the presence of a specific pathogen, these signals can be substances that leak from the lysed bacteria and can be picked up and read by different devices or chemical changes in the environment, which can be detected in different ways, indicating the bacteria's death (Chen et al. 2015). Among other ways, detection can also be done simply by viral expansion and subsequent increase in viral titer in the sample tested (Anany et al. 2017, Richter et al. 2018).

The next topics will generally address some applications for phagic cocktails in different settings, highlighting studies' results as well as discussing prospects and future possibilities for such applications.

2.1 *Bacteriophages in livestock and food production*

Bacteriophage versatility and selectivity against bacterial pathogens proportionated the expansion of these viruses in animal farms and food production facilities. Where the high usage of antibiotics increased the demand for alternative antibacterials that can both counter multidrug resistant (MDR) bacteria and avoid the emergence of new antibiotic-resistant strains (Ferriol-González and Pilar 2021).

Animal farm waste accounts for most of the "production" of ARGs, with cattle and fish wastewater presenting comparable ARGs concentrations with hospital and municipal wastes while swine and chicken wastes may present 3–5 higher orders of magnitude (He et al. 2020). Such higher ARG presence is likely a product of the elevated antibiotics application in animal farms, as this sector accounts for 50–80% of the total antibiotic application in developing countries (Cully 2014).

Another important point to highlight is that the presence of ARG in animal industries also promotes a constant human health hazard, since many of the antibiotics applied in animal production are also used in clinical settings (Aidara-Kane et al. 2018). Such factors may contribute to the fact that an estimated 60% of the emerging infectious diseases between 1940 and 2004 in humans were of zoonotic origin, with 54% of these events being caused by bacterial pathogens (Jones et al. 2008). Therefore, the indiscriminate usage of antibiotics in animal farms may consequently select clinical relevant MDR bacteria, being livestock health (and potential livestock targeting therapeutics) intrinsically intertwined with human health (Aidara-Kane et al. 2018, Ferriol-González and Pilar 2021).

Bacterial infections account for great losses in animal farms, causing animal diseases which may spread across an entire production chain, such as metritis, haemorrhagic septicaemia and mastitis (Santos et al. 2010, Qureshi et al. 2018, Ngassam-Tchamba et al. 2020). Alternatively, although it may not cause diseases in the livestock, asymptomatic bacterial infections, such as *Escherichia coli* O157:H7, may still contaminate food production chains and cause disease in humans, through certain asymptomatic strains (and even in the natural present microbiota) important targets to prevent zoonotic diseases (Sheng et al. 2006). In this sense, bacteriophage therapy was applied successfully against a large number of animal pathogens, whether by *in vivo* or *in vitro* assays, as demonstrated in Table 2.

The contamination control of the animal facility as a whole is necessary to avoid bacterial infection from the complex to the animals, which may arise by means of contaminated food, water, pens or even staff (Crump et al. 2002, Connor et al. 2017). Such a factor is clearly observed in fish farming plants, where water pathogens can easily propagate to the animals, for which phage treatment of fish ponds was able to control both water and animal bacterial load (Park and Nakai 2003).

Another characteristic of livestock infections for which bacteriophages provide a suitable approach, is the presence of persistent pathogenic bacteria, which are often unable to be controlled by antibiotics, thus remaining active in the animal host for the extent of their lifetime (Stevens et al. 2009). For example, *S. enterica* infections may persist in the host in a "carrier state" potentially perpetuating diseases in cattle, pigs and poultry (Stevens et al. 2009). Alternatively, *S. enterica* may persist without causing clinical signs but still actively replicate in the host, causing the animal to excrete infective bacterial loads and "silently" contaminate a food production chain (Stevens et al. 2009). The infective and self-replicative nature of bacteriophage therapy, was able to successfully infect and disrupt persistent bacteria in mammals, likely being the only available object in the treatment of persistent infections (Abedon 2019, Johri et al. 2021).

Table 2. Bacteriophage therapy in farm animals.

Bacterial pathogen	Disease	Animal	Phage treatment	References
E. coli	Metritis	Bovines	Reduction of isolates *in vitro*	(Santos et al. 2010)
Pasteurella multocida	Haemorrhagic Septicaemia	Bovines	Reduction of isolates *in vitro*	(Qureshi et al. 2018)
Aeromonas hydrophila	Haemorrhagic Septicaemia	Striped catfish (*Pangasianodon hypophthalmus*)	*In vivo* oral administration	(Dang et al. 2021)
Salmonella enterica serovar Typhimurium	Asymptomatic infection (zoonosis)	Swine	*In vivo* oral administration	(Wall et al. 2010, Seo et al. 2018)
Staphylococcus aureus	Bovine mastitis	Bovines	Reduction of isolates *in vitro*; treatment of *Galleria mellonella in vivo* model	(Ngassam-Tchamba et al. 2020)
Campylobacter jejuni	Asymptomatic infection (zoonosis)	Poultry	*In vivo* oral administration	(Loc-Carrillo et al. 2005)
E. coli O157:H7	Asymptomatic infection (zoonosis)	Bovines	*In vivo* oral administration; *in vivo* rectoanal administration	(Rivas et al. 2010, Sheng et al. 2006)
Staphylococcus aureus	Sinusitis	Sheep	*In vivo* intranasal administration	(Drilling et al. 2014)
Pseudomonas plecoglossicida	Bacterial hemorrhagic ascites	*Plecoglossus altivelis*	*In vivo* oral administration and water treatment	(Park and Nakai 2003)

Bacteriophages provide another factor to control systematic and hard-to-reach bacterial infections, in which their systemic auction may remove the need of topical administration in the infected tissue. Phage applications were observed to reach virtually all tissues, regardless of administration route, a characteristic signalized mainly by the phenomenon of transcytosis in which phage particles are able to cross cell barriers, even in complex tissues such as the blood-brain-barrier (Nguyen et al. 2017, Dabrowska and Abedon 2019). Such behavior confers phage-based therapies the potential ability to reach the infection of multiple organs, as was observed in the treatment of systemic *P. plecoglossicida* contaminated fish, for which orally fed phages were detected and able to control both intestine and kidney infections (Park and Nakai 2003).

Although animal phage treatment revealed promising results, failure on therapy reproductivity and variable success rate still raise concerns among specialists and companies, as therapy success may vary depending on the immunogenicity of the administration route, immunogenicity of the phage strains, distance between the targeted tissue and local of administration, phage titer, therapy exposure time and presence of hindering body factors like anti-phage antibodies (Dabrowska et al. 2014, Dabrowska and Abedon 2019, Łusiak-Szelachowska et al. 2014, Majewska et al. 2015, Schooley et al. 2017, Hodyra-Stefaniak et al. 2015).

Different phage strains, concentration and time after administration were seen influencing *C. jejuni* control in broiler chickens, varying the decrease in bacterial concentration from 0.5 to 5 log^10 CFU/g after the treatment (Loc-Carrillo et al. 2005). Remarkably, a higher phage concentration may exert a lesser effect than lower dosages, as was demonstrated in the same experiment, where log10[7] PFU decreased more *C. jejuni* in a 24 h interval than log10[9]. Failures in phage therapy after high titers inoculums were also observed in other studies, in which high phage concentrations were related to adaptive immunity induction on the animal against the viral therapy, resulting in the production of anti-phage antibodies which diminished phage activity and increased body clearance of the particles (Majewska et al. 2015, Srivastava et al. 2004, Hodyra-Stefaniak et al. 2015). Therefore, the mapping of the final phage product pharmacology is advisable for an optimal therapy result in live animals, for the body environmental factors may unexpectedly influence the therapy effect (Dadrowska 2019, Qadir et al. 2018).

The innate antibacterial nature of bacteriophages can also be potentially harmful to the animal's health if commensal/probiotic bacteria are directly targeted or disrupted by microbiota shifts. Such concern was expressed by several authors after the observations of microbiome shifts being directly related to phage activity, proportioning the development (or aggravation) of microbiome-tied diseases like Parkison's, type 1 diabetes and ulcerative colitis (Tetz et al. 2018b, Tetz et al. 2019c, Norman et al. 2015). Microbiome disruption (or dysbiosis) may happen even if the commensal bacteria is not directly targeted, as phage interference in the bacterial ecosystem may occasion unintentional upregulations or downregulations of non-targeted bacterial strains (Hsu et al. 2019, Tetz et al. 2017a).

Since animal microbiome balance is an important factor in cattle, pigs, poultry and fish health, the mapping of a phage product microbiome interactions is also highly desirable to avoid unintentional health alterations that may arise from the therapy (Zeineldin et al. 2018, Niederwerder 2017, Aruwa et al. 2021, Bozzi et al. 2021, Hsu et al. 2019).

2.2 Applications of phages in agriculture

The main goal of our century involves the development of sustainable chain-food-production, resulting in a decrease of pathogens/disease and healthy plants and animals (Ramankutty et al. 2018, Muller et al. 2017). The application of antibiotics aiming to control pathogens in crops cooperate to the emergence of antimicrobial resistance. *Erwinia amylovora* is an example of a pathogen responsible for diseases in apple and pear orchards which developed resistance to streptomycin (Schröpfer et al. 2021, Förster et al. 2015, Sholberg et al. 2001). The impact of antibiotic resistance has been discussed in scientific literature, considering the dynamics of our society (Timmerer et al. 2020, Minden et al. 2017).

In this context, bacteriophages became a useful tool that can be applied in agriculture. The first experiment with bacteriophages and plants demonstrated that a filter obtained from decomposing cabbage was capable of avoiding the cabbage-rot, caused by *Xanthomas campestris* pv. *campestris* (Mallmann and Hemstreet 1924). The first application on field was conducted during 1935, when Stewart's wilt disease of corn, disease caused by *Pantoea stewartii* was reduced when the seeds were pre-treated with specific phages (Katznelson 1937).

Two pathogens, *Pectobacterium* spp., and *Dickeya solani* are known to cause potato blackleg disease in the field and tuber soft rot in storage. Considering that, Adriaenssens applied *Dickeya* sp. *Myoviridae* phages aim to control the bacteria responsible for the disease. Phage preparation was applied as spray, at MOI of 10 over the infested potatoes. Treated potatoes were planted in the field, and monitored during their growth and development. Phage treatment resulted in 13% of yield and a decrease of symptoms as suppression of rotting (Adriaenssens et al. 2012).

Czajkowski and collaborators isolated 28 different tailed bacteriophages from soil, seeking the infection and lyse of 99 strains of *Dickeya* and *Pectobacterium* spp. Among these, they possess environmental strains: 8 isolates of *D. dadantii*, 6 isolates of *D. dianthicola*, 5 isolates of *D. zeae*, 2 isolates of *D. paradisiaca*, 4 isolates of *D. chrysanthemi*, 16 isolates of *D. solani*, 25 isolates of *P. atrosepticum*, 24 isolates of *P. carotovorum* subsp. *carotovorum*, 7 isolates of *P. wasabiae* and 2 isolates of *P. carotovorum* subsp. *brasiliense*. The screening resulted on two phages with capability to infect all strains: φPD10.3 and φPD23.1. These phages were evaluated against a high number of bacteria cells (5×10^5 CFU per inoculation), the protection was efficiency under 28°C and relative humidity (80%). These conditions were capable of reducing soft rot infections by at least 80% to 95% in comparison to controls inoculated with a mixture of bacteria only (Czajkowski et al. 2015).

Pseudomonas syringae is a pathogenic bacteria that is responsible for several plant diseases (Xin et al. 2018, Mansfield et al. 2012). An outbreak related first in Italy 2008 spread to other regions of Europe, resulting in massive infection of kiwi fruit by *Pseudomonas syringae* pv. *actinidiae*, causing economic loss of crops (Butler et al. 2013). This outbreak rekindled interest in the use of bacteriophages to biocontrol pathogenic bacteria (Frampton et al. 2014). In the light of those necessities, studies evaluating bacteriophages characterization by host range and genomic analyses were conducted. Trials were conducted by two parallel field applications, in three locations to the control of *P. syringae* pv. *porri*. The treatment consisted of a 6-phage cocktail, applied as pre-treatment to plants free of pathogens and plants treated after infection, the quantification of phages on cocktail was 10^9 pfu/ml. The results were not consistent, data suggested that this kind of application required optimization on formulation and application methods to be improved and efficient as biocontrol (Rombouts et al. 2016).

Balogh and collaborators evaluated the action of phage to biocontrol *Xanthomonas axonopodis* pv. *citri*, which causes the disease Asiatic Citrus Canker (ACC) and *X. axonopodis* pv. *citrumelo*, which causes the disease citrus bacterial spot (CBS). The copper-macozeb was used to control, during the experiment, phages were applied alone (1×10^9 PFU/ml) and combined with copper-macozeb in nursery trials. The effects of phage were not significantly different from copper-macozeb, and combinations of phage-copper-macozeb were not effective;

these results were observed against ACC and CBS (Balogh et al. 2008). In 2017, Ibrahim and collaborators found a methodology to improve the biocontrol of bacteriophages against *Xanthomonas axonopodis* pv. *citri*. The experiment was conducted in a greenhouse and field trials combining bacteriophages with a systemic acquired resistance inducer. The compounds utilized were acibenzolar-S-menthyl, which demonstrated a reduction of 82–86% in ACC incidence, emphasizing the importance of improvement methodologies of application/biocontrol using bacteriophages (Ibrahim et al. 2017).

Another pathogen with global relevance is *Xylella fastidiosa*. This bacteria possesses the capacity to cause diseases in several plants. In economic terms, when infected, grapes are the plant that cause the most losses, considering the production of wine (Chatterjee et al. 2008). Ahern and collaborators isolated and characterized bacteriophages from samples of plants and sewage with specificity to *X. fastidiosa* subsp. *fastidiosa*. As a result, the first characterization of bacteriophage applied to control of *X. fastidiosa* (Ahern et al. 2013).

Two years later, in 2015, a research was conducted evaluating the efficiency of four bacteriophages (Sano, Salvo, Prado and Paz) isolated by Ahern against *X. fastidiosa*. The application of cocktail (1×10^{10} PFU/ml) to plants was conducted as treatment and prophylactic. In the treatment the plants were inoculated with 1×10^9 CFU/ml. After 3 weeks, the cocktails were applied. As prophylaxi *X. fastidiosa* were applied after 3 weeks of inoculum of phages. The research demonstrated that it was efficient in both methods, resulting in no symptoms of disease after 6 weeks and 8 weeks respectively for each treatment (Das et al. 2015).

Apple and pears are fruits affected globally by the pathogen *Erwinia amylovora*, which causes the disease fire blight. The pathogens can be found on asymptomatic tissues or as an epiphyte in the orchard ecosystem (Tancos et al. 2017). In general, all the apple and pears cultivars spread throughout the world possess moderate to high susceptibility to this pathogen, and a resistant germplasm is not available yet. In Canada and the United States two antibiotics commonly chosen to treat this disease are streptomycin and kasugamycin (Mcmanus 2014, Mcmanus et al. 2002). Nevertheless, in some regions, bacteria present resistance to these drugs and/or the priority of production are crops free of antibiotics. This results in the necessity of alternatives. In this way, Boulé isolated and characterized eight bacteriophages that infect *Erwinia amylovora* and evaluated their capability of infection and biocontrol. Phage Ea2345-6 combined with Eh21-5 was capable of reducing around 56% of infection on apple flowers of potted apple trees, compared with use of antibiotics. Experiment utilized an MOI 1, being applied 1×10^8 CFU and PFU/ml, resulting in an alternative treatment to antibiotics (Boulé et al. 2011).

These studies demonstrate how the future of crops and agriculture in general can be optimized using bacteriophages as ally to combat pathogens, generating an increase of production of food and less contamination of soil and water and presence of resistance genes in bacteria.

2.3 *Applications of phages in the food industry*

Despite sanitization techniques and constant surveillance of pathogens, foodborne diseases remain one of the leading causes of hospitalization and death worldwide. According to the World Health Organization, about 600 million, one in ten people, get sick from eating contaminated food and about 420,000 people die annually (WHO 2020).

Several antimicrobial techniques were developed over time, such as pasteurization, high temperature processing, irradiation, and chemical disinfectants. However, they can present considerable disadvantages such as unwanted food cooking, loss of nutritional value, alteration of organoleptic properties, and corrosion of food processing equipment, respectively (Moye et al. 2018, Wolbang et al. 2008, Suklim et al. 2014, Sohaib et al. 2015).

Given this scenario, bacteriophages stand out and become a great alternative as a natural antimicrobial. The phage characteristics that make them promising include the high specificity for the host bacterium determined by receptors present on the bacterial cell wall, their simple composition of nucleic acids and proteins, low toxicity, high shelf life, resistance to environmental stresses in food processing, their self-replication and self-limitation means that few or single doses will multiply while there is still a host threshold present. Also, the phages can adapt to changes in the bacteria's defense mechanisms that occur over time (Sillankorva et al. 2012, Połaska and Sokołowska 2019).

Phages can be applied at different stages along the food chain to ensure food safety. The first recognized step for phage application, already explored in the previous topic, reduces the colonization of pathogenic bacteria in live animals. Subsequent steps concern disinfection of food contact surfaces and equipment and control of microbial growth in food processing. Finally, food preservation, through the prevention of contamination and proliferation of pathogens during the storage of final products (Sillankorva et al. 2012, Endersen and Coffey 2020).

Food, processing equipment, and surfaces can become contaminated by pathogenic bacteria such as *E. coli*, *Salmonella* sp., *Listeria monocytogenes,* and *Staphylococcus aureus* by contact with soil, water, fertilizers, humans, animals, and other contaminants (Verran et al. 2008, Abuladze et al. 2008). Possibly allowing the attachment of bacteria with the subsequent development of biofilms on contaminated areas. The presence of *Staphylococcus aureus* on food contact surfaces in dairy, meat, and seafood environments was analyzed in a study. The results highlight that surfaces of food industries can be reservoirs of *S. aureus*, aggravating the situation by the formation of complex multispecies biofilm communities with undesirable bacteria (Gutiérrez et al. 2012). Another study used different materials like viz plastic, cement, and stainless steel, to evaluate the biofilm formation and growth of two *Salmonella* isolates. The results show that biofilms have not been removed with the conventional cleaning procedure with sanitizers and therefore can also be sources of food contamination (Joseph et al. 2001).

In the disaggregation of biofilms, phages need to surpass the biofilm matrix. For this, phages can produce polysaccharide depolymerase that allows the invasion and dispersion through the biofilm (Barbirz et al. 2009). Beside the polysaccharide depolymerase, some phages produce lytic proteins such as endolysins and virion-associated peptidoglycan hydrolases (VAPGHs), also known as antimicrobial agents against pathogens. The VAPGHs are responsible for the "lysis from without", caused by the production of a small hole in the cell wall and adsorption of a high number of phages at the initial infection step (Moak and Molineux 2004). In contrast, the endolysins degrade the peptidoglycan layer of the host bacterium 'from within' at the end of their lytic multiplication cycle (Rodríguez-Rubio et al. 2015). Holins, the second component of the lysis cassette of tailed phages, are produced during the late stages of infection and can create holes in the cytoplasmic membrane, allowing the endolysins, which have accumulated in the cytoplasm, to access the peptidoglycan layer of the host bacterium (Schmelcher et al. 2012).

In addition to biosanitizing activity, phages can also be applied to reduce bacterial colonization in foods such as carcasses, raw products, and ready-to-eat (RTE) products (Sillankorva et al. 2012, García et al. 2008). One study highlighted a 3.0 log decrease of a *Salmonella* inoculum in chicken breast and milk after phage cocktail application (Islam et al. 2019). Another study was successful in applying a phage cocktail against *Shigella* to RTE, decreasing about 1.0–1.4 logs compared to the control (Soffer et al. 2017). There are also studies reporting phage cocktails against *Staphylococcus aureus*, capable of decreasing bacterial counts in cheeses without influencing its natural microbiota (Bueno et al. 2012).

Food safety becomes a leading challenge and concerns food industries when associated with stocking fresh-cut fruits, vegetables, and other fresh foods. Due to the high availability of nutrients on the surface of these foods, they become favorable to the colonization of pathogenic microorganisms and potential transmitters of foodborne diseases in humans. The challenge to solve the problems related to possible food contamination becomes even greater since fresh food cannot be preserved using traditional methods such as pasteurization, heat, and dehydration (Jebri 2020, Leverentz et al. 2003). Besides, the high demand for minimally processed foods and consumer awareness of potential health risks associated with traditional preservatives requires the development of new strategies for food preservation (Martínez et al. 2008). Once again, the phages become a promising alternative as a natural antimicrobial to reduce pathogenic bacteria in fresh food.

Studies have shown a decrease in the population of *E. coli* O157:H7 in fresh-cut lettuce and cantaloupes after being treated with a mixture of three bacteriophages and stored at 4°C (Sharma et al. 2009). A commercially available phage cocktail significantly reduced *L. monocytogenes* contamination in lettuce, cheese, smoked salmon, and frozen entrees (Perera et al. 2015).

Currently, several phage-based products are available on the market, showing the phage's potential in the food industry. Intralytix Inc. is a company that develops ListShield, EcoShield, and ShigaShield, products that target contamination in foods and food processing equipment by *L. monocytogenes, Salmonella* spp., and *Shigella* spp., respectively. SalmoFresh targets contamination by *Salmonella* spp. only in foods (INTRALYTIX INC.). The PhageGuard develop Listex, Salmonelex, and PhageGuard E targeting *L. monocytogenes, Salmonella* spp., and *E. coli*, respectively, in foods (PHAGEGUARD). These products are also USDA (United

States Department of Agriculture) and FDA (Food and Drug Administration) approved. However, the EFSA declared that it is not clear whether bacteriophages can protect food against re-contamination despite reports that bacteriophages are effective in the elimination of pathogens (EFSA 2012).

Although phages are very effective in targeting pathogenic bacteria in diverse environments, some challenges remain. Studies report that the levels of bacterial contamination drop soon after the application of bacteriophages, but then rise again (Guenther et al. 2012). A potential solution to this challenge is to use higher concentrations of phage particles to increase the likelihood that phages come into contact with their target bacteria after application, although this is a more expensive solution. Another option is to use larger spray volumes applied via fine mist sprays to disperse the phage particles across the food surface, increasing the likelihood of finding a target bacterium (Guenther et al. 2012, Kang et al. 2013). An important challenge related to the application of phage is the possibility of the development of resistant bacteria, which may eventually be selected by phage widespread usage. The use of phage cocktails may provide a mechanism to reduce the risk/probability of bacterial resistance (Soffer et al. 2017, Abuladze et al. 2008). Currently, consumers have shown greater interest in organic and preservative-free products, increasingly distancing themselves from products treated with chemical disinfectants and antibiotics or transgenics, giving space to phages. However, the idea of adding viruses to your food can be strange. Thus, it is necessary to inform the public about the safety, efficacy and ubiquity of bacteriophages (Naanwaab et al. 2014).

Despite the challenges, bacteriophages use has shown a safe and effective method. So far, the main applications of phages have been discussed. From their application in live animals to the storage of the final products. All the applications reported so far and the efforts made to develop new strategies to prevent or treat food contamination are related to food safety.

Food safety faces several challenges, such as the growing global population expected to reach 9.7 billion people by 2050 (FAO 2017), leading to increased demand for food and an overload of natural resources (Tilman et al. 2011). Besides, natural disasters create pathways for pathogens, chemicals, heavy metals, and other pollutants that can contaminate air, water, the environment, and food (Watson et al. 2007, Knorr et al. 2017, Andrade et al. 2018). Globalization and international travel make cross-border diseases a leading concern, as these diseases can sometimes be zoonotic and pose a risk to human health. The current century has created a new dynamic between people, animals, and animal products that affect food safety and security. This scenario requires a transformation of thinking and actions to face these contemporary challenges that threaten human health and well-being as well as animal and environmental health (King 2012). A holistic and systematic approach composed of academic, industry and government teams in a One Health strategy is logical and essential for success.

One Health approach addresses the ecological, animal, and environmental sources and influences needed to face the challenges outlined above. Actions in One Health include education programs in agriculture and food systems as an alternative for the involvement of the parties. The improvement of public health through activities should be supported by funding of innovative research and collaborations that provide new information and perspectives through food safety, and sustainable food production (Garcia et al. 2020). For continued progression in finding solutions, it is necessary to continue to use fundamental scientific research to inform regulations, practices and advanced technological applications to increase food production, improve sustainable practices and assess environmental impact. Furthermore, it is necessary to overcome the artificial separation between veterinary and animal health and public health. For example, in many zoonotic diseases, including foodborne diseases, the risks are associated with human health, while the most effective control strategies are in animals, animal products, and the environment (King 2012).

2.4 Phage application in wastewater

Water is a vital asset, essential for social, economic and energy development and food production. Despite being an undeniable right of any citizen, water scarcity resulting from social inequality and lack of management still places a large part of the world population at risk. The lack of potable water distribution systems forces about 1.2 billion people to consume water that did not undergo proper treatment, being exposed to pathogens related to the inappropriate dumping of waste in the waters. The consumption of contaminated food and water causes annually 1.5 million deaths of children under five years of age from diarrheal diseases. The lack of waste collection and treatment systems implies its inappropriate disposal in the environment, causing contamination of water resources. It is estimated that about 4.2 billion people, basically half of the world's population, do not have basic sanitation systems, evidencing the need to invest in the collection and treatment of wastewater to prevent the advancement of environmental contamination (WHO 2019).

Wastewater is water from the disposal of different processes. They can have a domestic character, such as water from baths, flushes and sinks; an industrial character, represented by the water derived from the manufacturing process of goods; an infiltration character, such as that collected by cisterns on land, or an urban character, which comes from rains, irrigation and washing of establishments (CETESB 2022). The inappropriate disposal of untreated wastewater harms the ecosystem and generates the potential for the emergence of disease outbreaks due to the presence of numerous chemical and biological contaminants. Many enteric pathogens, such as bacteria and viruses, are excreted at high rates in the feces of infected humans, making wastewater an important source of contamination. Recreational activities such as swimming, as well as shellfish consumption, also represent sources of risk of contagion in addition to the consumption of contaminated water and food (Nygård 2017, Prieto 2001).

To prevent the proliferation of these contaminants, the recommended process is that all kinds of wastewater should be collected and sent to a wastewater treatment plants (WWTP) in order to carry out the complete removal of contaminating materials, thus being finally able to deposit it into the environment (CETESB 2022). Among the steps of the treatment process is the removal of contaminants such as coliforms and some classes of pharmaceuticals. The treatment carried out in these plants is often insufficient to remove antibiotics, which ends up influencing the microbiota present in the wastewater and in the water source that will receive these waters after treatment. Some alternatives are being studied for application regarding the removal of antibiotics from wastewater, such as chlorination and UV treatment (Rivera-Utrilla et al. 2013).

The addition of chlorine as a decontaminant is already a common procedure in WWTP. However, as with other chemical substances present, chlorine reacts with some antibiotics, favoring its precipitation and accumulation in the sludge or giving rise to toxic compounds. The treatment that applies UV light emissions on water proved to be very effective in removing pathogens, including eliminating antibiotic-resistant bacteria, but UV photolysis of antibiotics often ends up forming more toxic and persistent compounds in the environment (Pinkston and Sedlak 2004, Tsamba et al. 2020). UV treatment is also not completely effective for disinfection by microorganisms, as some strains of *P. aeruginosa* have shown to be resistant both to the action of antibiotics naturally found in wastewater and to UV treatment (Rivera-Utrilla et al. 2013, Schwartz et al. 2015).

In the 20th century, phage therapy continued to be tested, producing phages to treat infections caused by a variety of bacterial species, such as *Staphylococcus*, *Pseudomonas*, *Proteus* and other enteric pathogens. Interest in the application of bacteriophages as bactericides has been renewed with the increasing number of antibiotic-resistant strains of bacteria that we have experienced in recent decades (Kortright et al. 2019). The development of studies on bacteriophages present in the environment evidenced the ubiquity and importance of these viral agents in the maintenance of ecosystems (Kutter and Sulakvelidze 2004). For this reason, several researches began to evaluate the potential of environmental applications of phages, using it to control bacterial infections in fish cultures, plants, animal waste, places of food preparation, carcasses and also preventing the blooms of cyanobacteria (Park and Nakai 2003, Flaherty et al. 2000, Mole et al. 1997, Thiel 2004). Another possible environmental application for the properties of bacteriophages is their use in wastewater treatment (Withey et al. 2005).

Bacteriophages may be a viable alternative for wastewater treatment, due to their natural ubiquity in the environment, specificity in infection and their role regulating the existing microbiota in all ecosystems (Jassim and Limoges 2014). Bacteriophages have already been isolated from different types of water sources with seawater being the environment with the highest density of bacteriophages present, reaching about 9×10^8 virions per milliliter of water (Wommack and Colwell 2000). Due to this high concentration of phages in seawater, it is estimated that about 70% of the bacteria present in the marine environment are already infected by bacteriophages (Jassim and Limoges 2013). After bacteria, phages are the most numerous category present in the microbiota of human feces, and phages of *Salmonella*, *E. coli* and *Bacteroides fragilis* serotypes have been isolated at concentrations of 10^5 plaque forming units (PFU)/10^{-2} g of fecal sample (Calci et al. 1998). Due to incorrect disposal of waste, bacteriophages of *Salmonella* have already been isolated in high concentrations in wastewater from all different sources (Havelaar 1987).

The use of bacteriophages as biological control through their bactericidal action has been studied for decades and their application in the treatment of wastewater with potential for the presence of pathogens has a number of related advantages. Bacteriophages represent an alternative to the use of chlorine and UV for the water treatment process, which can avoid changes in the physical-chemical parameters of the water and minimally harm the local microbiota. Another possible advantage is the ability of bacteriophages to replicate and increase their concentration in the medium to the point of meeting the demand for bacterial population control (Jassim and Limoges 2014).

The ability to break down biofilms present within the water treatment system also favors the use of phages in wastewater treatment (Jassim et al. 2012). The formation of biofilms in waste treatment systems is usually stimulated due to the mild temperatures of the water, the composition of the material used in the pipeline and the cycles of water flow, which periodically provide nutrients to the bacteria (Peres 2011). The species *P. aeruginosa*, for example, produces biofilms that commonly obstruct the passage of water in WWTP, requiring the application of high-cost measures, such as the use of chlorine and pressurized discharges. To look for more practical and low-cost ways to solve this issue, *P. aeruginosa* bacteriophages were isolated and applied to these biofilms, with results compared to those obtained in common chlorine treatments (Zhang and Hu 2013). Upon binding to the target bacterial cell, bacteriophages secrete the enzyme polysaccharide depolymerase, which degrades the exopolysaccharide capsule of the bacteria, reducing biofilm formation (Tait et al. 2002). While the chlorine treatment was able to remove only 40% of the biofilms, the application of phages at a concentration of 10^7 PFU removed 89% of the biofilms. These experiments suggest that the application of bacteriophages to kill bacteria can be effective on a variety of surfaces (Zhang and Hu 2013).

Bacteriophages do not have their action impaired by antibiotic-resistant bacteria, being an alternative of biological control for application in the treatment of wastewater from places with intensive use of these compounds, such as places where livestock is practiced, as well as hospitals and medical centers. The rampant use of antibiotics is generating an increasing amount of antibiotic-resistant strains of bacteria. Animal husbandry, for example, where significant amounts of antibiotics are distributed, represent a potential source of contaminants that, when reaching water sources, can end up harming the local microbiota, influencing the microbiota of the aquatic ecosystem (Jassim and Limoges 2014).

In addition to being potential disinfecting candidates capable of improving the efficiency of the wastewater treatment process, through the control of the population of bacteria that are excreted in the feces, bacteriophages also arouse interest regarding the ability to eliminate filamentous bacteria that impair the formation of activated sludge. Activated sludge is the result of an aerobic wastewater treatment process commonly performed in WWTP. To obtain it, the waste is placed in aerated tanks and added to a solution containing a community of microorganisms between autotrophic and heterotrophic bacteria, filamentous bacteria, algae, fungi, yeasts and protozoa. This microbiota will be responsible for consuming the organic matter from the wastewater, converting the available energy into biomass (Rustum 2009).

During this treatment process, the material suspended in the wastewater is slowly sedimented, forming sludge, and the supernatant is sent to the next stages of treatment. This sedimentation process is hampered by the replication of filamentous bacteria such as *Microthrix parvicella, Nostocoida limicola* and *Nocardia* spp. Characteristics of wastewater, such as substrate quality, pH and temperature, influence the dynamics of the activated sludge microbiota, with excess lipids and slowly decomposing organic materials especially favoring the growth of filamentous bacteria. Filamentous bacteria are capable of emitting long tentacles, in addition to dispersing gas bubbles, which end up adhering to the suspended material, preventing sedimentation. They are an important group for the formation of activated sludge, as they guarantee the sedimentation of solid materials, but the disordered growth of the population of bacteria ends up preventing the formation of a clear supernatant (Rustum 2009, Pal et al. 2014).

In order to solve this problem caused by filamentous bacteria, it was sought to add chlorine at this stage of the process, thus eliminating those responsible for impairing the sedimentation of the sludge. The addition of this disinfectant, however, eliminates not only the harmful bacteria that have replicated in excess, but also the beneficial ones, responsible for the degradation of organic matter, in addition to having its effect eliminated in a few hours, with the consequent return of the growth of the filamentous bacteria (Jassim et al. 2016). Experiments that tried to evaluate the potential of applying bacteriophages to control and eliminate these filamentous bacteria showed not only the formation of a clearer supernatant after 12 hours of phage application, but also a reduction in the volume of sludge produced. Regarding the durability and stability of the treatment, phages remain activated for nine months in the system, resisting the pH and temperature variations typical of the wastewater treatment process (Choi et al. 2011).

The research led by Grami and his team (2021) is an example where the application of bacteriophages to control the bacterial population of wastewater has been successful. The aim of their study was to isolate and describe bacteriophages from wastewater for further application in wastewater under different treatment levels. The bacteriophage PA25 was able to control and reduce the population of *P. aeruginosa* in both primary and secondary wastewater, being more efficient in those with less treatment process, indicating an advance in treatments based on the application of chlorine or UV emission, which require previous treatment steps to be effective. The action of PA25 in sterilized wastewater inoculated with *P. aeruginosa* ATCC strain 27853 was also evaluated, in comparison with non-sterile wastewater with the natural presence of the bacteria. Surprisingly, the bacteriophage was less efficient in controlling the bacterial population artificially inoculated in the sterilized water, both in the case of primary and secondary wastewater. This result suggests the importance of the interaction between bacteriophage and the wastewater bacterial community, since phages and bacteria are in constant competition in the environment, causing them to continually modify and adapt in relation to proteins and receptors. Due to the efficiency in wastewater disinfection pointed out by this study, bacteriophages were signalized as a potential tool in the control of wastewater bacterial populations under different treatment processes (Grami et al. 2021).

Despite the many advantages obtained from the application of bacteriophages as bactericides capable of eliminating species that are harmful to human and animal health, many researchers believe that there are still many challenges related to the use of phages on a large scale, as in WWTP. Some of the problems encountered are related to the number of groups of phages and the high concentration at which they need to be used in order to be effective against bacteria. The microbiota prevalent in different compositions of wastewater is very diversified, with the existence of more than 268 different species of bacteria in wastewater treatment systems being stipulated (Wagner and Loy 2002). Bacteria such as the *E. coli* and the *Salmonella* genus also have numerous serotypes present in wastewater and in the treatment system, with more than 2400 *Salmonella* serotypes already described (Fratamico et al. 2016, Grimont and Weill 2007). Applying a treatment with phages aiming to reach all species and serotypes present is unfeasible, making it necessary to use polyvalent phages (Withey et al. 2005), which presents a broad spectrum of hosts. However, the use of polyvalent phages in the wastewater treatment process could lead to the elimination of beneficial bacteria that promote the degradation of organic matter. Finally, the composition of the microbiota present in different WWTP varies, implying the necessity to analyze and describe the microorganisms present at the application site before the start of treatment (Khairnar et al. 2014, Withey et al. 2005).

Therefore, for large-scale application, further studies are needed to discover, develop and evaluate the efficiency of phages that eliminate only harmful bacteria. It is believed that such phages can be found by selecting groups that act only against harmful bacteria, maintaining the beneficial bacteria, decomposers of organic matter and other compounds that are difficult to degrade such as pharmaceuticals, without the need for genetic engineering. There are still several obstacles to be overcome so that bacteriophages can be applied *in situ* against pathogenic bacteria present in wastewater. However, due to its already proven efficiency and practicality, it may be the best alternative to offer supplies of drinking water completely free of any contaminants to the population (Khairnar et al. 2014, Withey et al. 2005).

The application of bacteriophages as bactericides in wastewater is presented as an ecological option that requires further studies so that it can be implemented on a large scale. Phage properties related to their density and ubiquity in the environment, mainly in the aquatic environment, as well as the absence of toxic metabolites, self-regulation in replication related to the size of the bacterial population, high specificity, which prevents them from harming the local microbiota and efficiency in the control of antibiotic-resistant bacteria suggest that they are organisms with great potential for use in the control of pathogenic bacteria present in wastewater. Strategies related to the characterization of the microbiota present in wastewater, as well as the identification of the bacteria-bacteriophage interaction in the aquatic environment, are essential for a successful use of this biocontrol method. With a greater understanding of the pathology and ecology of bacteria and bacteriophages in wastewater, it is possible to take advantage of the beneficial properties of phages (Pal 2014, Withey et al. 2005).

3. Future perspectives

This chapter introduced the reader to a variety of phage applications, in different settings and conditions that are currently being studied extensively. Despite all the biotechnological possibilities that these entities can provide, much remains to be understood, as well as many disadvantages that need to be resolved. In spite of the uncertainties that have been briefly pointed out in this chapter, phages are undoubtedly an important tool in bacterial control that will show tremendous potential in the future.

References

Abedon, S.T. 2019. Use of phage therapy to treat long-standing, persistent, or chronic bacterial infections. Advanced Drug Delivery Reviews 145: 18–39.

Abuladze, T., Li, M., Menetrez, M.Y., Dean, T., Senecal, A. and Sulakvelidze, A. 2008. Bacteriophages reduce experimental contamination of hard surfaces, tomato, spinach, broccoli, and ground beef by *Escherichia coli* O157:H7. Applied and Environmental Microbiology 74: 6230–6238.

Ackermann, H.W. 2009. Phage classification and characterization. Methods in Molecular Biology: 127–140.

Ackermann, H.W. and Prangishvili, D. 2012. Prokaryote viruses studied by electron microscopy. Archives of Virology 157: 1843–1849.

Adriaenssens, E.M., Van Vaerenbergh, J., Vandenheuvel, D., Dunon, V., Ceyssens, P.-J., De Proft, M. et al. 2012. T4-Related bacteriophage limestone isolates for the control of soft rot on potato caused by "Dickeya solani." PLoS ONE 7: 1–10.

Ahern, S.J., Das, M., Bhowmick, T.S., Young, R. and Gonzalez, C.F. 2013. Characterization of novel virulent broad-host-range phages of Xylella fastidiosa and Xanthomonas. Journal of Bacteriology 196: 459–471.

Aidara-Kane, A., Angulo, F.J., Conly, J.M., Minato, Y., Silbergeld, E.K., McEwen, S.A. and Collignon, P.J. 2018. World Health Organization (WHO) guidelines on use of medically important antimicrobials in food-producing animals. Antimicrobial Resistance & Infection Control 7: 1–8.

Altamirano, G.F.L. and Barr, J.J. 2019. Phage therapy in the postantibiotic era. Clinical Microbiology Reviews 32: 1–25.

Alves, D.R., Perez-Esteban, P., Kot, W., Bean, J., Arnot, T., Hansen, L., Enright, M.C. and Jenkins, A.T.A. 2016. Bacteriophages to treat *P. aeruginosa* biofilms. Microbial Biotechnology 9: 61–74. https://doi.org/10.1111/1751-7915.12316.

Anany, H., Brovko, L., El Dougdoug, N.K., Sohar, J., Fenn, H., Alasiri, N. et al. 2017. Print to detect: a rapid and ultrasensitive phage-based dipstick assay for foodborne pathogens. Analytical and Bioanalytical Chemistry 410: 1217–1230.

Andrade, L., O'Dwyer, J., O'Neill, E. and Hynds, P. 2018. Surface water flooding, groundwater contamination, and enteric disease in developed countries: a scoping review of connections and consequences. Environ. Pollut. 236: 540–549.

Aruwa, C.E., Pillay, C., Nyaga, M.M. and Sabiu, S. 2021. Poultry gut health—microbiome functions, environmental impacts, microbiome engineering and advancements in characterization technologies. Journal of Animal Science and Biotechnology 12: 1–15.

Balogh, B., Canteros, B.I., Stall, R.E. and Jones, J.B. 2008. Control of citrus canker and citrusbacterial spot with bacteriophages. Plant Dis. 92: 1048–1052.

Barbirz, S., Becker, M., Freiberg, A. and Seckler, R. 2009. Phage tailspike proteins with beta-solenoid fold as thermostable carbohydrate binding materials. Macromol Biosci. 9: 169–173.

Bozzi, D., Rasmussen, J.A., Carøe, C., Sveier, H., Nordøy, K., Gilbert, M.T.P. and Limborg, M.T. 2021. Salmon gut microbiota correlates with disease infection status: potential for monitoring health in farmed animals. Animal Microbiome 3: 1–17.

Boulé, J., Sholberg, P.L., Lehman, S.M., O'gorman, D.T. and Svircev, A.M. 2011. Isolation and characterization of eight bacteriophages infecting *Erwinia amylovora* and their potential as biological control agents in British Columbia, Canada. Canadian Journal of Plant Pathology 33: 308–317.

Brigati, J.R. and Petrenko, V.A. 2005. Thermostability of landscape phage probes. Analytical and Bioanalytical Chemistry 382: 1346–1350.

Bueno, E., García, P., Martínez, B. and Rodríguez, A. 2012. Phage inactivation of Staphylococcus aureus in fresh and hard-type cheeses. International Journal of Food Microbiology 158: 23–27.

Butler, M.I., Stockwell, P.A., Black, M.A., Day, R.C., Lamont, I.L. and Poulter, R.T.M. 2013. *Pseudomonas syringae* pv. actinidiae from recent outbreaks of kiwifruit bacterial canker belong to different clones that originated in China. PLoS ONE 8: 1–18.

Calci, K.R., Burkhardt III, W., Watkins, W.D. and Rippey, S.R. 1998. Occurrence of male-specific bacteriophage in feral and domestic animal wastes, human feces, and human-associated wastewaters. Applied and Environmental Microbiology 64: 5027–5029.

CETESB. 2022. São Paulo. Companhia Ambiental do Estado de São Paulo. Águas Interiores: Tipos de Águas. 2022. Available in: https://cetesb.sp.gov.br/aguas-interiores/informacoes-basicas/tpos-de-agua/.

Chatterjee, S., Almeida, R.P.P. and Lindow, S. 2008. Living in two Worlds: The Plant and Insect Lifestyles of Xylella fastidiosa. Annual Review of Phytopathology 46: 243–271.

Chen, J., Alcaine, S.D., Jiang, Z., Rotello, V.M. and Nugen, S.R. 2015. Detection of *Escherichia coli* in drinking water using T7 bacteriophage-conjugated magnetic probe. Analytical Chemistry 87: 8977–8984.

Choi, J., Kotay, S.M. and Goel, R. 2011. Bacteriophage-based biocontrol of biological sludge bulking in wastewater. Bioengineered Bugs 2: 214–217.

Clokie, M.R.J., Millard, A.D., Letarov, A.V. and Heaphy, S. 2011. Phages in nature. Bacteriophage 1: 31–45.

Costerton, J.W., Lewandowski, Z., Caldwell, D.E., Korber, D.R. and Lappin-Scott, H.M. 1995. Microbial biofilms. Annual Review of Microbiology 49: 712–739.

Crump, J.A., Griffin, P.M. and Angulo, F.J. 2002. Bacterial contamination of animal feed and its relationship to human foodborne illness. Clinical Infectious Diseases 35: 859–865.

Connor, J.T.O., Clegg, T.A. and More, S.J. 2017. Efficacy of washing and disinfection in cattle markets in Ireland. Irish Veterinary Journal 70: 1–6.

Cully, M. 2014. Public health: The politics of antibiotics. Nature 509: 16–17.

Czajkowski, R., Ozymko, Z., de Jager, V., Siwinska, J., Smolarska, A., Ossowicki, A. et al. 2015. Genomic, proteomic and morphological characterization of two novel broad host lytic bacteriophages ΦPD10.3 and ΦPD23.1 infecting Pectinolytic Pectobacterium spp. and Dickeya spp. PLOS ONE 10: 1–23.

Dabrowska, K., Miernikiewicz, P., Piotrowicz, A., Hodyra, K., Owczarek, B., Lecion, D. et al. 2014. Immunogenicity studies of proteins forming the T4 phage head surface. Journal of Virology 88: 12551–12557.

Dąbrowska, K. and Abedon, S.T. 2019. Pharmacologically aware phage therapy: pharmacodynamic and pharmacokinetic obstacles to phage antibacterial action in animal and human bodies. Microbiology and Molecular Biology Reviews 83: 1–45.

Dąbrowska, K. 2019. Phage therapy: What factors shape phage pharmacokinetics and bioavailability? Systematic and critical review. Medicinal Research Reviews 39: 2000–2025.

Dang, T.H.O., Xuan, T.T.T., Duyen, L.T.M., Le, N.P. and Hoang, H.A. 2021. Protective efficacy of phage PVN02 against haemorrhagic septicaemia in striped catfish Pangasianodon hypophthalmus via oral administration. Journal of Fish Diseases 44: 1255–1263.

Das, M., Bhowmick, T.S., Ahern, S.J., Young, R. and Gonzalez, C.F. 2015. Control of pierce's disease by phage. PLOS ONE 10: 1–15.

Davey, M.E. and O'toole, G.A. 2000. Microbial biofilms: from ecology to molecular genetics. Microbiology and Molecular Biology Reviews 64: 847–867.

Dion, M.B., Oechslin, F. and Moineau, S. 2020. Phage diversity, genomics and phylogeny. Nature Reviews Microbiology 18: 125–138.

D'Herelle, F. 1917. On an invisible microbe antagonistic toward dysenteric bacilli: brief note by Mr. F. D'Herelle, presented by Mr. Roux. Res. Microbiol. 158: 553–554.

Drilling, A., Morales, S., Boase, S., Jervis-Bardy, J., James, C., Jardeleza, C. et al. 2014. Safety and efficacy of topical bacteriophage and ethylenediaminetetra acetic acid treatment of *Staphylococcus aureus* infection in a sheep model of sinusitis. International Forum of Allergy & Rhinology 4: 176–186.

Duckworth, D.H. 1976. Who discovered bacteriophage? Bacteriol. Rev. 40: 793–802.

Edgar, R., Friedman, N., Molshanski-Mor, S. and Qimron, U. 2012. Reversing bacterial resistance to antibiotics by phage-mediated delivery of dominant sensitive genes. Applied and Environmental Microbiology 78: 744–751.

EFSA. 2012. Scientific Opinion on the evaluation of the safety and efficacy of ListexTM P100 for the removal of *Listeria monocytogenes* surface contamination of raw fish. EFSA Journal 10: 1–43.

Endersen, L. and Coffey, A. 2020. The use of bacteriophages for food safety. Current Opinion in Food Science 36: 1–8.

FAO. 2017. The Future of Food and Agriculture—Trends and Challenges. Food and Agriculture Organization of the United Nations. Rome.

Ferriol-González, C. and Pilar, D.C. 2021. Phage therapy in livestock and companion animals. Antibiotics 10: 1–12.

Flaherty, J.E., Somodi, G.C., Jones, J.B., Harbaugh, B.K. and Jackson, L.E. 2000. Control of bacterial spot on tomato in the greenhouse and field with h-mutant bacteriophages, HortScience HortSci 35: 882–884.

Förster, H., McGhee, G.C., Sundin, G.W. and Adaskaveg, J.E. 2015. Characterization of streptomycin resistance in isolates of Erwinia amylovora in California. Phytopathology® 105: 1302–1310.

Frampton, R.A., Taylor, C., Holguín Moreno, A.V., Visnovsky, S.B., Petty, N.K., Pitman, A.R. et al. 2014. Identification of bacteriophages for biocontrol of the kiwifruit canker phytopathogen *Pseudomonas syringae* pv. *actinidiae*. Applied and Environmental Microbiology 80: 2216–2228.

Fratamico, P.M., DebRoy, C., Liu, Y., Needleman, D.S., Baranzoni, G.M. and Feng, P. 2016. Advances in molecular serotyping and subtyping of *Escherichia coli*. Frontiers in Microbiology 7: 1–8.

Fu, W., Forster, T., Mayer, O., Curtin, J.J., Lehman, S.M. and Donlan, R.M. 2010. Bacteriophage cocktail for the prevention of biofilm formation by *Pseudomonas aeruginosa* on catheters in an *in vitro* model system. Antimicrobial Agents and Chemotherapy 54: 397–404.

García, P., Martínez, B., Obeso, J.M. and Rodríguez, A. 2008. Bacteriophages and their application in food safety. Letters in Applied Microbiology 47: 479–485.

Garcia, S.N., Osburn, B.I. and Jay-Russell, M.T. 2020. One health for food safety, food security, and sustainable food production. Frontiers in Sustainable Food Systems 4: 1–9.

Grami, E., Salhi, N., Sealey, K.S., Hafiane, A., Ouzari, H.I. and Saidi, N. 2021. Siphoviridae bacteriophage treatment to reduce abundance and antibiotic resistance of *Pseudomonas aeruginosa* in wastewater. International Journal of Environmental Science and Technology 19: 3145–3154.

Grimont, P.A. and Weill, F.X. 2007. Antigenic formulae of the Salmonella serovars. WHO Collaborating Centre for Reference and Research on Salmonella 9: 1–166.

Gu, J., Liu, X., Li, Y., Han, W., Lei, L., Yang, Y. et al. 2012. A method for generation phage cocktail with great therapeutic potential. PLoS ONE 7: 1–8.

Guenther, S., Herzig, O., Fieseler, L., Klumpp, J. and Loessner, M.J. 2012. Biocontrol of Salmonella Typhimurium in RTE foods with the virulent bacteriophage FO1-E2. International Journal of Food Microbiology 154: 66–72.

Gutiérrez, D., Delgado, S., Vázquez-Sánchez, D., Martínez, B., Cabo, M.L., Rodríguez, A. et al. 2012. Incidence of *Staphylococcus aureus* and analysis of associated bacterial communities on food industry surfaces. Applied and Environmental Microbiology 78: 8547–8554.

Harper, D., Parracho, H., Walker, J., Sharp, R., Hughes, G., Werthén, M., Lehman, S. and Morales, S. 2014. Bacteriophages and biofilms. Antibiotics 3: 270–284.

Havelaar, A.H. 1987. Virus, bacteriophages and water purification, Veterinary Quarterly 9: 356–360.

He, Y., Yuan, Q., Mathieu, J., Stadler, L., Senehi, N., Sun, R. and Alvarez, P.J.J. 2020. Antibiotic resistance genes from livestock waste: occurrence, dissemination, and treatment. Npj Clean Water 3: 1–11.

Hodyra-Stefaniak, K., Miernikiewicz, P., Drapała, J., Drab, M., Jończyk-Matysiak, E., Lecion, D. et al. 2015. Mammalian Host-Versus-Phage immune response determines phage fate *in vivo*. Scientific Reports 5: 1–13.

Hsu, B.B., Gibson, T.E., Yeliseyev, V., Liu, Q., Lyon, L., Bry, L. et al. 2019. Dynamic modulation of the gut microbiota and metabolome by bacteriophages in a mouse model. Cell Host & Microbe 25: 803–814.

Ibrahim, Y.E., Saleh, A.A. and Al-Saleh, M.A. 2017. Management of asiatic citrus canker under field conditions in Saudi Arabia using bacteriophages and acibenzolar-s-methyl. Plant Disease 101: 761–765.

Islam, Md.S., Zhou, Y., Liang, L., Nime, I., Liu, K., Yan, T. et al. 2019. Application of a phage cocktail for control of salmonella in foods and reducing biofilms. Viruses 11: 1–19.

Jassim, S.A.A., Abdulamir, A.S. and Abu Bakar, F. 2011. Novel phage-based bio-processing of pathogenic *Escherichia coli* and its biofilms. World Journal of Microbiology and Biotechnology 28: 47–60.

Jassim, S.A., Abdulamir, A.S. and Abu Bakar, F. 2012. Novel phage-based bio-processing of pathogenic *Escherichia coli* and its biofilms. World J. Microbiol. Biotechnol. 2012 Jan; 28(1): 47–60. doi: 10.1007/s11274-011-0791-6. Epub 2011 May 22. PMID: 22806779.

Jassim, S.A.A. and Limoges, R.G. 2013. Impact of external forces on cyanophage–host interactions in aquatic ecosystems. World Journal of Microbiology and Biotechnology 29: 1751–1762.

Jassim, S.A. and Limoges, R.G. 2014. Natural solution to antibiotic resistance: bacteriophages 'The Living Drugs'. World Journal of Microbiology & Biotechnology 30: 2153–2170.

Jassim, S.A.A., Limoges, R.G. and El-Cheikh, H. 2016. Bacteriophage biocontrol in wastewater treatment. World Journal of Microbiology and Biotechnology 32: 1–10.

Jebri, S. 2020. Bacteriophages as preservation agents to promote minimally processed food safety. Nutrition and Food Science International Journal 10: 1–2.

Johri, A.V., Johri, P., Hoyle, N., Pipia, L., Nadareishvili, L. and Nizharadze, D. 2021. Case report: chronic bacterial prostatitis treated with phage therapy after multiple failed antibiotic treatments. Frontiers in Pharmacology 12: 1–8.

Jones, K.E., Patel, N.G., Levy, M.A., Storeygard, A., Balk, D., Gittleman, J.L. et al. 2008. Global trends in emerging infectious diseases. Nature 451: 990–993.

Joseph, B., Otta, S.K., Karunasagar, I. and Karunasagar, I. 2001. Biofilm formation by Salmonella spp. on food contact surfaces and their sensitivity to sanitizers. International Journal of Food Microbiology 64: 367–372.

Khairnar, K., Pal, P., Chandekar, R.H. and Paunikar, W.N. 2014. Isolation and characterization of bacteriophages infecting nocardioforms in wastewater treatment plant. Biotechnology Research International 2014: 1–5.

Kang, H.W., Kim, J.W., Jung, T.S. and Woo, G.J. 2013. wksl3, a New Biocontrol Agent for *Salmonella enterica* Serovars Enteritidis and Typhimurium in foods: characterization, application, sequence analysis, and oral acute toxicity study. Applied and Environmental Microbiology 79: 1956–1968.

Katznelson, H. 1937. Bacteriophage in relation to plant diseases. Botanical Review 3: 499–521.

King, L.J. 2012. One Health and Food Safety. National Academies Press, Washington.

Knorr, W., Dentener, F., Lamarque, J.-F., Jiang, L. and Arneth, A. 2017. Wildfire air pollution hazard during the 21st century. Atmos. Chem. Phys. 17: 9223–9236.

Kortright, K.E., Chan, B.K., Koff, J.L. and Turner, P.E. 2019. Phage therapy: a renewed approach to combat antibiotic-resistant bacteria. Cell Host & Microbe 25: 219–232.

Kutter, E. and Sulakvelidze, A. 2004. Bacteriophages: Biology and Applications. CRC Press, Florida.

Leverentz, B., Conway, W.S., Camp, M.J., Janisiewicz, W.J., Abuladze, T., Yang, M. et al. 2003. Biocontrol of Listeria monocytogenes on fresh-cut produce by treatment with lytic bacteriophages and a bacteriocin. Applied and Environmental Microbiology 69: 4519–4526.

Lin, D.M, Koskella B. and Lin, H.C. 2017. Phage therapy: An alternative to antibiotics in the age of multi-drug resistance. World J. Gastrointest Pharmacol. Ther. 8: 162–173.

Loc-Carrillo, C., Atterbury, R.J., El-Shibiny, A., Connerton, P.L., Dillon, E., Scott, A. and Connerton, I.F. 2005. Bacteriophage therapy to reduce Campylobacter jejuni colonization of broiler chickens. Applied and Environmental Microbiology 71: 6554–6563.

Łusiak-Szelachowska, M., Żaczek, M., Weber-Dąbrowska, B., Międzybrodzki, R., Kłak, M., Fortuna, W. et al. 2014. Phage neutralization by sera of patients receiving phage therapy. Viral Immunology 27: 295–304.

Martínez, B., Obeso, J.M., Rodríguez, A. and García, P. 2008. Nisin-bacteriophage crossresistance in *Staphylococcus aureus*. International Journal of Food Microbiology 122: 253–258.

Majewska, J., Beta, W., Lecion, D., Hodyra-Stefaniak, K., Kłopot, A., Kaźmierczak, Z. et al. 2015. Oral application of T4 phage induces weak antibody production in the gut and in the blood. Viruses 7: 4783–4799.

Mallmann, W. and Hemstreet, C. 1924. Isolation of an inhibitory substance from plants. Journal of Agricultural Research 28: 1–4.

Mansfield, J., Genin, S., Magori, S., Citovsky, V., Sriariyanum, M., Ronald, P. et al. 2012. Top 10 plant pathogenic bacteria in molecular plant pathology. Molecular Plant Pathology 13: 614–629.

Martinez-Hernandez, F., Fornas, O., Gomez, M.L., Bolduc, B., Peña, M.J.D.L.C., Martínez, J.M. et al. 2017. Single-virus genomics reveals hidden cosmopolitan and abundant viruses. Nature Communications 8: 1–13.

McManus, P.S., Stockwell, V.O., Sundin, G.W. and Jones, A.L. 2002. Antibiotic use in plant agriculture. Annual Review of Phytopathology 40: 443–465.

McManus, P.S. 2014. Does a drop in the bucket make a splash? Assessing the impact of antibiotic use on plants. Current Opinion in Microbiology 19: 76–82.

Minden, V., Deloy, A., Volkert, A.M., Leonhardt, S.D. and Pufal, G. 2017. Antibiotics impact plant traits, even at small concentrations. AoB PLANTS 9: 1–19.

Moak, M. and Molineux, I.J. 2004. Peptidoglycan hydrolytic activities associated with bacteriophage virions. Mol. Microbiol. 51: 1169–1183.

Mole, R., Meredith, D. and Adams, D.G. 1997. Growth and phage resistance of Anabaena sp. strain PCC 7120 in the presence of cyanophage AN-15. Journal of Applied Phycology 9: 339–345. https://doi. org/10.1023/A:1007938624025.

Moye, Z., Woolston, J. and Sulakvelidze, A. 2018. Bacteriophage applications for food production and processing. Viruses 10: 205.

Muller, A., Schader, C., El-Hage Scialabba, N., Brüggemann, J., Isensee, A., Erb, K.-H. et al. 2017. Strategies for feeding the world more sustainably with organic agriculture. Nature Communications 8: 1–13.

Naanwaab, C., Yeboah, O.A., Ofori Kyei, F., Sulakvelidze, A. and Goktepe, I. 2014. Evaluation of consumers' perception and willingness to pay for bacteriophage treated fresh produce. Bacteriophage 4: 1–7.

Ngassam-Tchamba, J.N., Duprez, M., Fergestad, A., De Visscher, T., L'Abee-Lund, S., De Vliegher, Y., Wasteson, F., Touzain, Y., Blanchard, R., Lavigne, N., Chanishvili, D., Cassart, J. and Mainil, D. Thiry. 2020. *In vitro* and *in vivo* assessment of phage therapy against *Staphylococcus aureus* causing bovine mastitis. Journal of Global Antimicrobial Resistance 22: 762–770.

Nguyen, S., Baker, K., Padman, B.S., Patwa, R., Dunstan, R.A., Weston, T.A., Schlosser, K., Bailey, B., Lithgow, T., Lazarou, M., Luque, A., Rohwer, F., Blumberg, R.S. and Barr, J.J. 2017. Bacteriophage transcytosis provides a mechanism to cross epithelial cell layers. mBio 8: 1–14.

Niederwerder, M.C. 2017. Role of the microbiome in swine respiratory disease. Veterinary Microbiology 209: 97–106.

Norman, J.M., Handley, S.A., Baldridge, M.T., Droit, L., Liu, C.Y., Keller et al. 2015. Disease-specific alterations in the enteric virome in inflammatory bowel disease. Cell 160: 447–460.

Nygård, K.M. 2017. Norwegian Institute of Public Health. Food and waterborne diseases. Public Health Report Editorial Group.

Octavia, S. and Lan, R. 2014. The Family Enterobacteriaceae. *In*: Rosenberg, E., DeLong, E.F., Lory, S., Stackebrandt, E. and Thompson, F. (eds.). The Prokaryotes. Springer, Berlin.

O'Sullivan, L., Bolton, D., McAuliffe, O. and Coffey, A. 2019. Bacteriophages in food applications: from foe to friend. Annual Review of Food Science and Technology 10: 151–172.

Pal, P., Khairnar, K. and Paunikar, W.N. 2014. Causes and remedies for filamentous foaming in activated sludge treatment plant. Global Nest J. 16: 762–772.

Park, S. and Nakai, T. 2003. Bacteriophage control of Pseudomonas plecoglossicida infection in ayu, Plectoglossis altivelis. Diseases of Aquatic Organisms 53: 33–39.

Pearl, S., Gabay, C., Kishony, R., Oppenheim, A. and Balaban, N.Q. 2008. Nongenetic individuality in the host-phage interaction. PLoS Biol. 5: 1–8.

Pelzek, A.J., Schuch, R., Schmitz, J.E. and Fischetti, V.A. 2008. Isolation, culture, and characterization of bacteriophages. Current Protocols Essential Laboratory Techniques 7(1): 4.4.1–4.4.33. doi:10.1002/9780470089941.et0404s07.

Perera, M.N., Abuladze, T., Li, M., Woolston, J. and Sulakvelidze, A. 2015. Bacteriophage cocktail significantly reduces or eliminates *Listeria monocytogenes* contamination on lettuce, apples, cheese, smoked salmon and frozen foods. Food Microbiology 52: 42–48.

Peres, B.M. 2011. Bactérias indicadoras e patogênicas em biofilmes de sistemas de tratamento de água, sistemas contaminados e esgoto. Dissertation. Universidade de São Paulo, São Paulo, Brasil.

Pinkston, K.E. and Sedlak, D.L. 2004. Transformation of aromatic ether- and amine-containing pharmaceuticals during chlorine disinfection. Environmental Science & Technology 38: 4019–4025.

Połaska, M. and Sokołowska, B. 2019. Bacteriophages—a new hope or a huge problem in the food industry. AIMS Microbiology 5: 324–346.

Prieto, M.D. 2001. Recreation in coastal waters: health risks associated with bathing in sea water. Journal of Epidemiology & Community Health 55: 442–447.

Principi, N., Silvestri, E. and Esposito, S. 2019. Advantages and limitations of bacteriophages for the treatment of bacterial infections. Frontiers in Pharmacology 10: 1–9.

Qadir, M.I., Mobeen, T. and Masood, A. 2018. Phage therapy: progress in pharmacokinetics. Brazilian Journal of Pharmaceutical Sciences 54: 1–9.

Qureshi, S., Saxena, H.M., Imam, N., Kashoo, Z., Sharief Banday, M., Alam, A. et al. 2018. Isolation and genome analysis of a lytic Pasteurella multocida Bacteriophage PMP-GAD-IND. Letters in Applied Microbiology 67: 244–253.

Ramankutty, N., Mehrabi, Z., Waha, K., Jarvis, L., Kremen, C., Herrero, M. et al. 2018. Trends in global agricultural land use: implications for environmental health and food security. Annual Review of Plant Biology 69: 789–815.

Ronayne, E.A., Wan, Y.C.S., Boudreau, B.A., Landick, R. and Cox, M.M. 2016. P1 Ref Endonuclease: A molecular mechanism for phage-enhanced antibiotic lethality. PLOS Genetics 12: 1–31.

Richter, Ł., Janczuk-Richter, M., Niedziółka-Jönsson, J., Paczesny, J. and Hołyst, R. 2018. Recent advances in bacteriophage-based methods for bacteria detection. Drug Discovery Today 23: 448–455.

Rivas, L., Coffey, B., McAuliffe, O., McDonnell, M.J., Burgess, C.M., Coffey, A., Ross, R.P. and Duffy, G. 2010. *In vivo* and *ex vivo* evaluations of bacteriophages e11/2 and e4/1c for use in the control of *Escherichia coli* O157:H7. Applied and Environmental Microbiology 76: 7210–7216.

Rivera-Utrilla, J., Sánchez-Polo, M., Ferro-García, M.Á., Prados-Joya, G. and Ocampo-Pérez, R. 2013. Pharmaceuticals as emerging contaminants and their removal from water. A review. Chemosphere 93: 1268–1287.

Rodríguez-Rubio, L., Gutiérrez, D., Donovan, D.M., Martínez, B., Rodríguez, A. and García, P. 2015. Phage lytic proteins: biotechnological applications beyond clinical antimicrobials. Critical Reviews in Biotechnology 36: 1–11.

Rombouts, S., Volckaert, A., Venneman, S., Declercq, B., Vandenheuvel, D., Allonsius, C.N. et al. 2016. Characterization of novel bacteriophages for biocontrol of bacterial blight in leek caused by *Pseudomonas syringae* pv. *porri*. Frontiers in Microbiology 7: 1–15.

Rustum, R. 2009. Modelling Activated Sludge Wastewater Treatment Plants Using Artificial Intelligence Techniques: Fuzzy Logic and Neural Networks. Ph.D Thesis. Heriot-Watt University, Edinburgh, Scotland.

Salmond, G. and Fineran, P. 2015. A century of the phage: past, present and future. Nat. Rev. Microbiol. 13: 777–786.

Santos, S.B. and Azeredo, J. 2019. Bacteriophage-based biotechnological applications. Viruses 11: 1–4.

Santos, T.M.A., Gilbert, R.O., Caixeta, L.S., Machado, V.S., Teixeira, L.M. and Bicalho, R.C. 2010. Susceptibility of *Escherichia coli* isolated from uteri of postpartum dairy cows to antibiotic and environmental bacteriophages. Part II: *In vitro* antimicrobial activity evaluation of a bacteriophage cocktail and several antibiotics. Journal of Dairy Science 93: 105–114.

Schmelcher, M., Donovan, D.M. and Loessner, M.J. 2012. Bacteriophage endolysins as novel antimicrobials. Future Microbiol. 7: 1147–1171.

Schooley, R.T., Biswas, B., Gill, J.J., Hernandez-Morales, A., Lancaster, J., Lessor, L. et al. 2017. Development and use of personalized bacteriophage-based therapeutic cocktails to treat a patient with a disseminated resistant Acinetobacter baumannii infection. Antimicrobial Agents and Chemotherapy 61: 1–30.

Schröpfer, S., Vogt, I., Broggini, G.A.L., Dahl, A., Richter, K., Hanke, M.-V. et al. 2021. Transcriptional profile of AvrRpt2EA-mediated resistance and susceptibility response to Erwinia amylovora in apple. Scientific Reports 11: 1–14.

Schwartz, T., Armant, O., Bretschneider, N., Hahn, A., Kirchen, S., Seifert, M. and Dötsch, A. 2015. Whole genome and transcriptome analyses of environmental antibiotic sensitive and multi-resistant *Pseudomonas aeruginosa* isolates exposed to waste water and tap water. Microbial Biotechnology 8: 116–130.

Seed, K., Lazinski, D., Calderwood, S., Camilli, A. et al. 2013. A bacteriophage encodes its own CRISPR/Cas adaptive response to evade host innate immunity. Nature 494: 489–491.

Seo, B.J., Song, E.T., Lee, K., Kim, J.W., Jeong, C.G., Moon, S.H. et al. 2018. Evaluation of the broad-spectrum lytic capability of bacteriophage cocktails against various *Salmonella* serovars and their effects on weaned pigs infected with *Salmonella* Typhimurium. Journal of Veterinary Medical Science 80: 851–860.

Sharma, M., Patel, J.R., Conway, W.S., Ferguson, S. and Sulakvelidze, A. 2009. Effectiveness of bacteriophages in reducing *Escherichia coli* O157:H7 on fresh-cut cantaloupes and lettuce†. Journal of Food Protection 72: 1481–1485.

Sheng, H., Knecht, H.J., Kudva, I.T. and Hovde, C.J. 2006. Application of bacteriophages to control intestinal *Escherichia coli* O157:H7 levels in ruminants. Applied and Environmental Microbiology 72(8): 5359–66. doi: 10.1128/AEM.00099-06. PMID: 16885287; PMCID: PMC1538718.

Sholberg, P.L., Bedford, K.E., Haag, P. and Randall, P. 2001. Survey of Erwinia amylovoraisolates from British Columbia for resistance to bactericides and virulence on apple. Canadian Journal of Plant Pathology 23: 60–67.

Sillankorva, S.M., Oliveira, H. and Azeredo, J. 2012. Bacteriophages and their role in food safety. International Journal of Microbiology 2012: 1–13.

Sohaib, M., Anjum, F.M., Arshad, M.S. and Rahman, U.U. 2015. Postharvest intervention technologies for safety enhancement of meat and meat based products: a critical review. Journal of Food Science and Technology 53: 19–30.

Soffer, N., Woolston, J., Li, M., Das, C. and Sulakvelidze, A. 2017. Bacteriophage preparation lytic for shigella significantly reduces shigella sonnei contamination in various foods. Plos One 12: 1–11.

Srivastava, A.S., Kaido, T. and Carrier, E. 2004. Immunological factors that affect the *in vivo* fate of T7 phage in the mouse. Journal of Virological Methods 115: 99–104.

Stevens, M.P., Humphrey, T.J. and Maskell, D.J. 2009. Molecular insights into farm animal and zoonotic Salmonella infections. Philosophical Transactions of the Royal Society B: Biological Sciences 364: 2709–2723.

Suklim, K., Flick, G.J. and Vichitphan, K. 2014. Effects of gamma irradiation on the physical and sensory quality and inactivation of *Listeria monocytogenes* in blue swimming crab meat (Portunas pelagicus). Radiation Physics and Chemistry 103: 22–26.

Suttle, C. 2005. Viruses in the sea. Nature 437: 356–361.

Suttle, C. 2007. Marine viruses—major players in the global ecosystem. Nature Reviews Microbiology 5: 801–812.

Tait, K., Skillman, L.C. and Sutherland, I.W. 2002. The efficacy of bacteriophage as a method of biofilm eradication. Biofouling 18: 305–311.

Tancos, K.A., Borejsza-Wysocka, E., Kuehne, S., Breth, D. and Cox, K.D. 2017. Fire blight symptomatic shoots and the presence of Erwinia amylovora in asymptomatic apple budwood. Plant Disease 101: 186–191.

Tanji, Y., Shimada, T., Yoichi, M., Miyanaga, K., Hori, K. and Unno, H. 2004. Toward rational control of *Escherichia coli* O157:H7 by a phage cocktail. Applied Microbiology and Biotechnology 64: 270–274.

Tetz, G.V., Ruggles, K.V., Zhou, H., Heguy, A., Tsirigos, A. and Tetz, V. 2017a. Bacteriophages as potential new mammalian pathogens. Scientific Reports 7: 1–9.

Tetz, G., Brown, S.M., Hao, Y. and Tetz, V. 2018b. Parkinson's disease and bacteriophages as its overlooked contributors. Scientific Reports 8: 1–11.

Tetz, G., Brown, S.M., Hao, Y. and Tetz, V. 2019c. Type 1 diabetes: an association between autoimmunity, the dynamics of gut amyloid-producing *E. coli* and their phages. Scientific Reports 9: 1–11.

Thiel, K. 2004. Old dogma, new tricks—21st Century phage therapy. Nature Biotechnology 22(1): 31–36. DOI: 10.1038/nbt0104-31.

Tilman, D., Balzer, C., Hill, J. and Befort, B.L. 2011. Global food demand and the sustainable intensification of agriculture. Proc. Natl. Acad. Sci. U.S.A. 108: 20260–20264.

Timmerer, U., Lehmann, L., Schnug, E. and Bloem, E. 2020. Toxic effects of single antibiotics and antibiotics in combination on germination and growth of Sinapis alba L. Plants 9: 1–19.

Tsamba, L., Correc, O. and Couzinet, A. 2020. Chlorination by-products in indoor swimming pools: Development of a pilot pool unit and impact of operating parameters. Environment International 137: 1–10.

Twort, F.W. 1915. An investigation on the nature of ultra-microscopic viruses. Lancet 2: 1241–1243.

Verran, J., Airey, P., Packer, A. and Whitehead, K.A. 2008. Chapter 8 Microbial retention on open food contact surfaces and implications for food contamination. Advances in Applied Microbiology 64: 223–246.

Xin, X.-F., Kvitko, B. and He, S.Y. 2018. Pseudomonas syringae: what it takes to be a pathogen. Nature Reviews Microbiology 16: 316–328.

Wall, S.K., Zhang, J., Rostagno, M.H. and Ebner, P.D. 2010. Phage therapy to reduce preprocessing Salmonella infections in market-weight swine. Appl. Environ. Microbiol. 2010 Jan; 76(1): 48–53. doi: 10.1128/AEM.00785-09. Epub 2009 Oct 23. PMID: 19854929; PMCID: PMC2798657.

Watson, J.T., Gayer, M. and Connolly, M.A. 2007. Epidemics after natural disasters. Emerg. Infect. Dis. 13: 1–5.

Wagner, M. and Loy, A. 2002. Bacterial community composition and function in sewage treatment systems. Current Opinion in Biotechnology 13: 218–227.

Weber-Dąbrowska, B., Jończyk-Matysiak, E., Żaczek, M., Łobocka, M., Łusiak-Szelachowska, M. and Górski, A. 2016. Bacteriophage procurement for therapeutic purposes. Frontiers in Microbiology 7: 1–14.

Withey, S., Cartmell, E., Avery, L.M. and Stephenson, T. 2005. Bacteriophages—potential for application in wastewater treatment processes. Science of The Total Environment 339: 1–18.

Wolbang, C.M., Fitos, J.L. and Treeby, M.T. 2008. The effect of high pressure processing on nutritional value and quality attributes of Cucumis melo L. Innovative Food Science & Emerging Technologies 9: 196–200.

Wommack, K.E. and Colwell, R.R. 2000. Virioplankton: viruses in aquatic ecosystems. Microbiology and Molecular Biology Reviews 64: 69–114.

World Health Organization. 2019. Sanitation: health topics. Health Topics. Available in: https://www.who.int/news-room/fact-sheets/detail/sanitation.

World Health Organization. 2020. Food safety. Who.int. Available in: https://www.who.int/news-room/fact-sheets/detail/food-safety.

Yacoby, I., Bar, H. and Benhar, I. 2007. Targeted drug-carrying bacteriophages as antibacterial nanomedicines. Antimicrobial Agents and Chemotherapy 51: 2156–2163.

Ye, M., Sun, M., Huang, D., Zhang, Z., Zhang, H., Zhang, S., Hu, F. et al. 2019. A review of bacteriophage therapy for pathogenic bacteria inactivation in the soil environment. Environment International 129: 488–496.

Yilmaz, C., Colak, M., Yilmaz, B.C., Ersoz, G., Kutateladze, M. and Gozlugol, M. 2013. Bacteriophage therapy in implant-related infections: An experimental study. J. Bone Jt. Surg. Am 95: 117–125.

Zeineldin, M., Aldridge, B. and Lowe, J. 2018. Dysbiosis of the fecal microbiota in feedlot cattle with hemorrhagic diarrhea. Microbial Pathogenesis 115: 123–130.

Zhang, Y. and Hu, Z. 2013. Combined treatment of *Pseudomonas aeruginosa* biofilms with bacteriophages and chlorine. Biotechnol. Bioeng. 110: 286–295.

Chapter 5

Detection of Enteric Viruses in Foods and Food-Processing Environments

Rachel Siqueira de Queiroz Simões[1,2,3] and
David Rodríguez-Lázaro[2,3,]*

1. Introduction

The promotion of a high level of food safety is a major policy worldwide. Moreover, the guarantee of safety and quality of food products along the food chain is the principal demand of the consumer as they expect their food to be tasty and wholesome as well as safe. The concern of safety of food products has been increased considerably during the last decades by the rapid globalisation of the food market, and profound changes in the food consumption habits. Foodborne virus infections are of great importance to public health being one of the global food safety-related priority areas of the World Health Organization (WHO) (Pexara and Govaris 2020). As mentioned in previous chapters, most of the viral foodborne illnesses are caused by human noroviruses and Hepatitis A virus, and other emerging viruses suchas Hepatitis E virus area attracting much attention in the food safety area. In addition, some emerging encephalitis-producing viruses such as flavivirus or Nipah virus can be also transmitted via the fecal-oral route transmission by manipulation or

---

[1] Institute of Technology in Immunobiologicals, Bio-Manguinhos, Oswaldo Cruz Foundation, Fiocruz, Avenida Brasil, 4365, Manguinhos, Rio de Janeiro 21040-900, Brazil.
[2] Microbiology Division, Faculty of Sciences, University of Burgos, 09001 Burgos, Spain.
[3] Research Centre for Emerging pathogens and Global Health, University of Burgos, 09001 Burgos, Spain.
* Corresponding author: drlazaro@ubu.es

consumption of contaminated dairy or pig meat products (Bosch et al. 2018, Pexara and Govaris 2020). The problem of foodborne virus infections has been also greatly increased by the effect of radical changes in the food market—the globalization of the food market—and in consumption habits as well as changes in consumer tastes.

Consequently, the implementation of microbiological quality control programs and surveillance systems that include the quantification of food-borne diseases and the identification of emerging pathogens is a premise to minimize the risk of infection for the consumer. Therefore, the development and implementation of new alternatives for the monitoring, characterization and enumeration of foodborne pathogenic viruses is one of the key aspects in food microbiology and has become a fundamental aspect for the food industry. However, detection of viruses in food and food-related environments is challenging due to the large variety and complexity of samples, the possible heterogeneous distribution of a small number of viruses and the presence of components that may inhibit or interfere with virus detection (Rodríguez-Lázaro et al. 2012).

During the last decades, different diagnostic approaches have been developed that have contributed significantly to this field. Several different approaches can be used to detect human enteric viruses in food samples, including direct observation by electron microscopy, detection of cytopathic effects in specific cell lines or detection using immunological or molecular methods. Regardless of the method used, detection of the presence of enteric viruses in foods is an important issue in food safety and requires rapid and robust diagnostic methodology (Bosch et al. 2011). In this chapter, the most important aspects in virus diagnostics in food and food-related environments will be reviewed.

2. Traditional and cut-of-the-edge detection methods for viruses in food

Traditional methods have been applied in virus detection in food (such as electron microscopy, cell culture, and serology). However, viral characteristics impair the effective implementation of these detection approaches. Usually, the viral particles are in few numbers in foods and diluted in large volumes of samples, requiring concentration methods before analysis (Rodríguez-Lázaro et al. 2012).

2.1 Direct observation by microscopy

Electron Microscopy has been used as a method to detect enteric virus particles in faecal specimens as well as in food since the early 70's. However, direct observation by electron microscopy needs a large number of viruses for detection and is time-consuming, subjective and of limited sensitivity (Atmar and Estes 2001). Generally, the suspension of biological specimens from samples collected in the first two days of symptoms are analyzed by transmission electron microscopy with a sensitivity of 10^5 to 10^6 particles/mL.

2.2 Cell culture-based method

The use of cell culture allows the replication of viruses in the laboratory and can be used for several purposes such as diagnosis-isolation of particular viruses-, or for the production of vaccines. When possible, some cell culture-based techniques such as Plaque or $TCID_{50}$ assays can be used. Plaque assays can enumerate plaque-forming viruses and is rather proportional to the number of virus particles, while the $TCID_{50}$ assay is based on the observation of cytopathic effects produced in specific cell lines as the modification of the morphology of the infected cell. However, some animal and human biochemical and physiological factors can interfere with viral biosynthesis and limit the isolation of specific viruses. The observation of cytopathic effects produced in specific cell lines is not always possible as some enteric viruses, notably human noroviruses and hepatitis E virus, cannot be propagated in mammalian cell lines (Chapron et al. 2000, Lipp et al. 2001). Even when it is possible, the detection of viruses using a cell culture is not a simple or cost-effective technique. This technique may also require the adaptation of the virus in order for it to grow effectively (Pintó and Bosch 2008).

2.3 Immunological methods

Immunological tests do exist, and many are commercially available for the main enteric viruses. However, their analytical sensitivity is still too poor for effective testing of food samples. Enzyme immunoassays such as ELISA are widely used in food microbiology. The system involves detection of the immunecomplex fixed on a support using monoclonal and polyclonal antibodies conjugated to an enzyme to capture the target antigen. The captured antigen is detected using a second antibody conjugated with an enzyme (e.g., peroxidase or alkaline phosphatase). To facilitate visualization of the target antigen, a substrate is added to produce a colorimetric reaction and the test result is determined by the spectrophotometric measurement produced by the reaction of the enzyme on the substrate. This assay has been applied in clinical samples for detection of foodborne viruses such as rotaviruses and adenoviruses. Another immunological assay is the reverse Passive Latex Agglutination (RPLA). This method is used for both thermosensitive and thermostable detection of microbial toxins. The latex particles are coated with rabbit serum antibody and react in the presence of the target antigen. The particles will agglutinate in the presence of the antigen forming a compact structure.

2.4 Molecular-based methods

To overcome the above mentioned limitations, molecular techniques and particularly quantitative real-time PCR (q-PCR) have become the method of choice for the detection of enteric viruses. This approach has been reinforced by the introduction of the international standard method ISO 15216-1:2017 for detection of human noroviruses and hepatitis A virus. The features of the ideal molecular method for detection of foodborne viruses are presented in Table 1. Molecular methods have the advantage over previously described tests of being able to analyze multiple samples simultaneously using automated procedures. One of the major differences between

Table 1. Main attributes for an ideal test for virus detection in food.

Internal (performance) characteristics
Selective (specific/inclusive and sensitive/exclusive)
With a low limit of detection and with a low limit of quantification (when enumeration is required)
Robust (accurate and precise)
External characteristics
Rapid
Versatile (ample range of matrices that can be tested, and/or ample range of viruses to be tested)
Economically effective (cost per food sample and initial investment for the associated instrumentation/equipment)
Automated (reduced number of handling steps and personnel involved)

the study of the presence and enumeration of bacteria and that of viruses in food and in the environment is the availability of a "gold standard" method for detection. Classical culture-based techniques are considered the gold standard for the detection of bacteria, but the situation is exactly the opposite for the detection of viruses, since no accepted standard method exists. The lack of a defined and consensus standard method for detection and quantification of viruses is hindering and slowing the adaptation of quantitative viral risk assessment (QVRA) models for food and food environments.

When a methodology is applied for virus detection in food, several steps are needed prior to final detection using molecular methods. It is necessary to separate and concentrate viruses from foods before performing detection assays, and establishing appropriate separation and concentration processes is very demanding. Whatever the method used, the final concentrate should be free of any inhibitors which may be co-extracted or co-concentrated from environmental samples (Goyal 2006). A variety of biological and chemical substances which are present in foods or are used during sample processing have been found to act as inhibitors, including polysaccharides, haeme, phenol, and cations (Atmar 2006). Known PCR inhibitors in shellfish extracts include glycogen and acidic polysaccharides (Schwab et al. 1998).

2.4.1 Matrix separation

Virus extraction from food is the first step for the detection of enteric viruses in food matrices. During this step, viral particles are separated from the food matrix, and the final concentrate must be free of any inhibitors which may be co-extracted or co-concentrated from the sample matrix (Stals et al. 2012). The extraction method highly depends on the composition of the food. Protocols for concentration of viruses in food samples often start with a washing step (in the case of fresh produce) or a homogenization step (in the case of, for example, shellfish); the virus is concentrated subsequent to this first step (Croci et al. 2008, Rodríguez-Lázaro et al. 2007). For this step, three main approaches can be followed: (i) elution of the viral particles with subsequent concentration; (ii) direct extraction of the viral RNA from the food matrix without the elution–concentration step; and (iii) extraction of viruses from the food by proteinase K treatment. After most of extraction protocols, purification (removal of food debris and inhibitory substances) of the virus eluate/concentrate or extracted RNA is performed to remove inhibitory compounds (e.g., polysaccharides, proteins, fatty acids), which could inhibit the subsequent nucleic acid purification and molecular detection (Stals et al. 2012).

The elution–concentration step is based on washing the viral particles from the food surface using an appropriate buffer followed by concentration of the eluted viruses. Several elution buffers have been used to elute viral agents from food matrices (Stals et al. 2012). Alkaline buffer at pH 9 to 10.5 is applied as an alkaline environment allows virus particles to detach from the food solids. The acidic nature of fruits and vegetables can impair virus elution and consequently reduce detection sensitivity (Dubois et al. 2002). After viruses had been extracted from fruits and vegetables, pectinase is frequently added to prevent jelly formation in the eluate by breaking the pectin bonds (Dubois et al. 2002). A neutral buffer for virus elution is mainly used when anionic exchange or ultracentrifugation are applied to concentrate the eluted viral particles. For foods which contain acidic substances (such as fruits and vegetables) an alkaline Tris-based buffer system is usually used (reviewed in Stals et al. 2012). Phosphate buffer (Bidawid et al. 2000a) or sodium bicarbonate buffer (Kurdziel et al. 2001) have been also used successfully. Elution buffers are frequently combined with beef extract (1–3%) and glycine, which reduce non-specific virus adsorption to the food matrix during their extraction (Dubois et al. 2002, Love et al. 2008) and can therefore increase virus recovery (Butot et al. 2007).

Direct virus nucleic acid extraction includes treatment of the food with guanidinium isothiocyanate (GITC)/phenol, followed by purification of the viral RNA. It has been used for norovirus concentration from foods (Boxman et al. 2007, Baert et al. 2008, Stals et al. 2011b). Direct RNA extraction combined with shredding of the digestive tissue using zirconium beads has successfully recovered noroviruses from oyster digestive tissue (de Roda Husman et al. 2007).

The combination of proteinase K and heat treatment (usually 65°C) damages the virus capsid and releases virus nucleic acid (Nuanualsuwan and Cliver 2002). This approach has been used successfully for the detection of noroviruses in several foodborne outbreaks related to shellfish (Jothikumar et al. 2005, Comelli et al. 2008) and has been selected by the ISO method 15216-1:2017 for the extraction of the most common enteropathogenic viruses from shellfish digestive tissue.

2.4.2 Concentration of viruses

The aim of concentrating viruses is to collect most of the viruses present in the sample in a minimal volume (Cliver 2008). Virus concentration methods appropriate for a wide variety of matrices include differential precipitation (using polyethylene glycol and Sodium Chloride in concentrations 0,5 M), ultracentrifugation (in high speed to remove cells debris and other residues from the food samples), ultrafiltration and adsorption-elution (viruses adsorb on the principle of ionic charge and elution by buffer to a small volume) (Rodríguez-Lázaro et al. 2007).

Viruses can also be concentrated by cationic separation or by immunoconcentration. Cationic separation uses positively charged magnetic particles in conjunction with a magnetic capture system to concentrate and purify viral particles from food matrices, as the negatively charged proteins of the virus capsid bind to the positively charged magnetic particles. This approach has been successfully used for concentration of noroviruses in lettuce and fresh cheese providing a recovery efficiency ranging from 5.2 to 56.3% (Fumian et al. 2009). Immunoconcentration methods can use magnetic beads covered with histo-blood group antigens (HBGA) or porcine gastric mucin for concentration of noroviruses (Tian et al. 2008, Morton et al. 2009) or magnetic beads covered by monoclonal hepatitis A virus antibodies (Bidawid et al. 2000a). This allows specific extraction of foodborne viruses in different categories of foods and efficient removal of PCR inhibitors. However, due to immunogenetic drift, the long term use of the same antibodies could be problematic (Stals et al. 2012).

2.4.3 Extraction of viral nucleic acids

Genomic nucleic acids must be first extracted from the viral concentrates using in-house or commercial procedures (Mattison and Bidawid 2009). A wide variety of commercial kits are currently available, most of them based on a method which uses guanidinium isothyocianate-based lysis followed by capture of nucleic acids on silica (Boom et al. 1990, Haramoto et al. 2005, Rutjes et al. 2005, Le Guyader et al. 2009). The use of commercial kits offers ease of use, reliability, reproducibility. In recent years, automated nucleic acid extraction platforms have been developed and have been efficiently applied for the analysis of viruses in food samples (Comelli et al. 2008, Perrele et al. 2009, Stals et al. 2011b). Other methods for viral nucleic acid extraction include proteinase K treatment followed by phenol-chloroform extraction and ethanol precipitation, sonication and heat treatment (Jothikumar et al. 2005, Guévremont et al. 2006, Comelli et al. 2008, Le Guyader et al. 2009).

2.4.4 Molecular-based detection

Conventional polymerase chain reaction (PCR) was the first molecular technique used for the detection of viral genomes. The sensitivity and specificity can be improved using semi-nested or nested PCR and they use an internal primer or primer set (Van Heerden et al. 2003). Similarly, Reverse transcription PCR (RT-PCR) has been reported as gold standard in detection of foodborne RNA viruses, and numerous detection kits have been developed. Real-time PCR (q-PCR) has become the preferable approach for detection of viruses in food. This assay has improved nucleic acid detection and became the preferable approach for detection of viruses in food. The development of the qPCR represents a significant advance in molecular techniques involving nucleic acids analysis. q-PCR allows monitoring of the synthesis of new amplicon molecules during the PCR (i.e., in real time). Data is therefore collected throughout the PCR process and not just at the end of the reaction (as occurs in conventional PCR) (Heid et al. 1996). Major advantages of q-PCR are the closed-tube format that avoids the risk of carryover contamination, fast and easy to perform analysis, the extremely wide dynamic range of quantification (more than eight orders of magnitude), and the significantly higher reliability and sensitivity of the results compared to conventional PCR (Rodríguez-Lázaro et al. 2007). However, q-PCR also suffers from some limitations. The volume used in the amplification reaction is very small; therefore only concentration methods that can deliver a very small volume of the resulting nucleic acid solution from a realistic food sample can be used. In addition, the quality of the nucleic acids is an important factor that directly affects the analytical sensitivity of the assay, and diverse compounds present in samples can inhibit the amplification reaction.

One of the principal drawbacks of molecular methods and particularly qPCR for detection of enteric viruses in food is the impossibility of distinguish between infective and non-infective viral particles. Virus infectivity is defined as the capacity of viruses to enter the host cell and exploit its resources to replicate and produce progeny infectious viral particles (Rodríguez et al. 2009), which may lead to infection and subsequent disease in the human host. Therefore, the key information from a public health perspective is the number of virus particles with infective capacity present in food. Obviously, cell-culture based methods are the soundest methodologies for the estimation of the number of infective particles. However, as indicated above, there are no available culture models for the most significant enteric viruses. This limits its usefulness for public health purposes. The ratio between genome equivalents (GE) and infectious particles has been reported to increase with the time, and is strongly dependent upon water and climatic conditions and virus type. Damage to the virus capsid may result in the loss of its capacity to protect the genome and its ability to replicate in the host. Consequently, the detection of an intact viral genome can be an indication that the virus capsid is still in good condition, protecting the genome from degradation. Determining the relationship between damage to the viral capsid and degradation of the viral genome can provide information that can be used to correlate the detection of the viral genome with the infectivity of the virus. Therefore, several strategies have been developed to adapt PCR to quantify infective virus particles (Rodriguez et al. 2009). Two different approaches have been used: direct RT-PCR (Ma et al. 1994, Li et al. 2002, Simonet and Gantzer 2006a, 2006b) or the use a pre-PCR sample treatment coupled to a subsequent RT-PCR (Nuanualsuwan and Cliver 2002, 2003).

Other molecular techniques mainly used in the detection of enteric viruses are Digital RT-PCR, Nucleic Acid Sequence-Based Amplification (NASBA) and loop-mediated isothermal amplification (LAMP). Digital RT-PCR (dRTPCR) is an endpoint quantitative approach that estimates genome copies based on the Poisson distribution. In terms of accuracy assay, the sensitivity of RTdPCR is comparable and upper to RTqPCR. Another advantage which this approach does nOt necessary quantitative data with standard curve (Bosh et al. 2016). Other molecular techniques mainly used in the detection of enteric viruses are Nucleic Acid Sequence-Based Amplification (NASBA) and loop-mediated isothermal amplification (LAMP). NASBA employs three different enzymes: T7 bacteriophage RNA polymerase, RNase H and avian myeloblastosis virus reverse transcriptase (AMV-RT), which act in concert to amplify sequences from an original single-stranded RNA template (Compton 1991). The reaction also includes two oligonucleotide primers, complementary to the RNA region of interest, and one of the primers also contains a promoter sequence that is recognized by T7 RNA polymerase at the 5'-end. LAMP has been used for detection of several foodborne viruses. At least four primers are used for amplification: two of which are the loop primers that recognize two regions each in the target genetic sequence. The target sequence is amplified using a strand-displacing DNA-dependent DNA polymerase, and detection is accomplished by measuring an increase in turbidity or the binding of a fluorescent detection reagent; this can be monitored in real-time (Fukuda et al. 2007, Yoneyama et al. 2007). The specificity of the reaction in the LAMP system can be increased using additional primers (Mattison and Bidawid 2009).

2.5 Next-generation sequencing and metagenomics

The detection and characterization of viruses in fecal samples, vomit and shellfish has been carried out through sequencing the genome of norovirus and HAV. News technologies in the field of "omics" are making high progress and will also be helpful in analysis of food samples. The metagenomics studies allowed the expansion of knowledge of new viral species providing new viral variation and briefly will contribute in the identification of the virome environmental samples in near future (Wylie et al. 2013, Bosh et al. 2018). These approaches are precious tools both in clinical and food studies. Metagenomics tools may be applied to different kinds of samples for viral studies (Hanza and Bibby 2019, Wang et al. 2016). More recently, metagenomic studies have allowed the discovery of new viral species, assisting in the understanding of viral communities and their biological role. The use of viral metagenomics has been applied for monitoring and diagnosis and viruses, included in food (Nieuwenhuijse and Koopmans 2017), being largely effective in molecular identification and traceability of these etiological agents. Viral metagenomics has caused an important revolution in virology, especially in environmental virology, revealing that viruses contained in the biosphere are not well represented by viruses purified and replicated in cell cultures *in vitro*, being important in the discovery and understanding of the virus-cell relationship.

NGS methodologies have been widely useful for knowing emerging viruses with potential for spread by water and food route, and the possibility to understand viral adaptation from wild to human origin. In addition, the metagenomics is interesting for environmental monitoring purposes, it is a promising solution since several viruses can be studied in a single sequencing run, being largely important for sanitary food control. Given the metagenomics perspectives of viral disease traceability and their origins, mainly food and water sources, metadata can be correlated, identified, evaluated and mitigated for risk reduction purposes. The food metagenomics approach performed by Hellmér et al. (2014) revealed viral multispecies transmitted by water and food routes, demonstrating that norovirus and hepatitis A found in food, for example, were related to outbreaks of hospitalized patients diagnosed with a viral infection. In studies conducted by Aw et al. (2014) analyzed viruses from human sewage, detecting various viruses of the family *Polyomaviridae*, *Picornaviridae* and *Papillomaviridae*. Similarly, viruses have been identified by metagenomics from fresh fruits and vegetables (Aw et al. 2014), dried meats and seafood to reduce risks (Severi et al. 2015). Metagenomic sequencing has also been used to monitor illegal meat (Temmam et al. 2016) in shellfish, demonstrating the accumulation of norovirus, sapovirus, and hepatitis A virus (Benabbes et al. 2013).

3. Characteristics of an ideal method for detection of viruses in food

It is a multifaceted task to define clearly what are the main characteristics, the first step is to understand what the major needs in the food supply chain are, to define the key features that an ideal (and never fully achievable) method must possess.

The first identified main need is a high (absolute) **selectivity**. Conceptually, selectivity is defined as "*a measure of the degree of non-interference in the presence of non-target analytes*" (ISO 2003). A method is selective if it can detect the given target virus under examination, guaranteeing that the detection signal is only generated by that specific virus. In other words, a method must always detect the given target virus when it is present –specificity– and not detect it when it is not present–sensitivity. This is particularly relevant in the food industry, since as a first step it must comply with domestic and/or international food regulations.

A second identified main need is the capacity for testing an **ample range of matrices**, in other words, the capacity of a method to be horizontal and applicable to different types of food categories such as plant products, meat and meat products, milk and dairy products, etc. This aspect is also very relevant as both food industry and food contract labs can reduce the arsenal of methods to be used, and make the process of accreditation for those tests less of an effort.

Another relevant need is a short time to **final confirmed results**, the shorter, the better. This is a relevant aspect, which has been clearly addressed by the food industry.

Another relevant need is to minimize the **cost of the analysis**. This can be dissected in two different aspects: routine costs (costs associated with the test kits and consumables and the costs associated with personnel) and capital investment costs (costs associated with instruments for analysis, in the case of molecular methods the PCR and/or qPCR platforms and associated equipment).

Finally, **automation** has also been defined as an essential attribute for the food industry. Time is an extremely valuable commodity and all the processes should be in-line. The analysis should therefore be rapid and easy to perform, and if possible non-destructive and carried out in an appropriate flow chain (in-line analysis). It is clear that it is not always possible, but obtaining of results in parallel with the production is a priority for the food industry. As mentioned above, the speed of obtaining final results is critical; simplification of the method, reduction of handling by personnel, and lab automation can facilitate this.

Once the major needs in the food chain are identified, it is much easier to define the expectations required for a detection method for viruses in food. These expectations must at the least encompass the fulfillment of the needs defined above. These can be clustered in two big groups: internal attributes and external attributes. The first group is related to characteristics associated with the performance of the method, while the second one is related to characteristics of the method that can facilitate effective implementation but without a direct effect on the analytical performance.

Finally, a very important characteristic is cost efficiency. In other microbiological areas, such as clinical or pharmaceutical microbiology diagnostics, cost is a relatively less important attribute, in comparison to those related to analytical performance.

Table 2. The features of the ideal procedure used for extraction and detection of foodborne viruses.

	Features	Comments
Extraction and concentration method	generic	concentrate vast range of foodborne viruses in a small volume of concentrate
	efficient	possess good virus recovery
	labor intensity	not time consuming, short procedure
	feasibility	it has to be easy to perform and can be done in suitably equipped virological laboratories
	robust	good reproducibility and repeatability
Detection method	detection limit	should be as low as possible allowing to detect only a few virus particles in food samples

However, the final cost of food products and the low profit margins that the food industry have, makes the control of any extra costs a main priority. Therefore, as the food industry is becoming more conscious of the importance and relevance of food safety, making it a primary priority, the cost of analysis of particular higher cost methodologies hinders their effective implementation in routine food labs. As an example, the cost of testing an RTE salad lot for foodborne viruses (i.e., testing 5 samples for 3 main viruses; human Noroviruses GI and GII and hepatitis A virus) can exceed 1,500 €. Assuming a conservative definition of an RTE salad lot is the quantity produced in 1 day (typically exceeding 1,000 RTE salad bags), the cost can represent from 0.05 to 0.5 € per RTE salad bag. This additional cost will either increase the price of the final product or reduce the company's profit.

In conclusion, the main features of the ideal procedure used for detection of foodborne viruses are summarised in Table 2. The focus is on maximizing virus recovery while minimizing loss due to the procedure. Future work should also be focused on development of extraction procedures which remove or minimize inhibitors. Another challenge is the issue of the infectivity of the detected viruses.

References

Atmar, R.L. and Estes, M.K. 2001. Diagnosis of noncultivatable gastroenteritis viruses, the human caliciviruses. Clin. Microbiol. Rev. 14: 15–37.

Aw, T.G., Howe, A. and Rose, J.B. 2014. Metagenomic approaches for direct and cell culture evaluation of the virological quality of wastewater. J. Virol. Methods 210: 15–21.

Baert, L., Uyttendaele, M. and Debevere, J. 2008. Evaluation of viral extraction methods on a broad range of ready-to-eat foods with conventional and real-time RT-PCR for Norovirus GII detection. International Journal of Food Microbiology 123(1-2): 101–108

Benabbes, L., Ollivier, J., Schaeffer, J., Parnaudeau, S., Rhaissi, H., Nourlil, J. et al. 2013. Norovirus and other human enteric viruses in moroccan shellfish. Food Environ. Virol. 5: 35–40. doi: 10.1007/s12560-012-9095-8.

Bidawid, S., Farber, J.M. and Sattar, S.A. 2000a. Rapid concentration and detection of hepatitis A virus from lettuce and strawberries. Journal of Virological Methods 88(2): 175–185

Boom, R., Sol, C.J., Salimans, M.M., Jansen, C.L., Wertheim-van Dillen, P.M. and van der Noordaa, J. 1990. Rapid and simple method for purification of nucleic acids. J. Clin. Microbiol. 28: 495–503.

Bosch, A., Sanchez, G., Abbaszadegan, M., Carducci, A., Guix, S., Le Guyader, F.S., Netshikweta, R., Pintó, R.M., van der Poel, W., Rutjes, S., Sano, D., Rodríguez-Lázaro, D., Kovac, K., Taylor, M.B., van Zyl, W. and Sellwood, J. 2011. Analytical methods for virus detection in water and food. Food Anal. Methods 4: 4–13.

Bosch, A., Pinto, R.M, and Guix, S. 2016. Foodborne viroses. Current Opinion in Food Science 8: 110–119.

Bosch, A. et al. 2018. Foodborne viruses: Detection, risk assessment, and control options in food processing. International Journal of Food Microbiology 285: 110–128.

Boxman, I.L., Tilburg, J.J., te Loeke, N.A., Vennema, H., de Boer, E. and Koopmans, M. 2007. An efficient and rapid method for recovery of norovirus from food associated with outbreaks of gastroenteritis. Journal of Food Protection 70(2): 504–508

Butot, S., Putallaz, T. and Sánchez, G. 2007. Procedure for rapid concentration and detection of enteric viruses from berries and vegetables. Applied and Environmental Microbiology 73(1): 186–192.

Cliver, D.O. 2008. Historic Overview of Food Virology. pp. 1–28. Koopmans, M.P.G., Cliver, D.O. and Bosch, A. (eds.). Foodborne Viruses: Progress and Challenges. ASM press, Washington, DC.

Comelli, H.L., Rimstad, E., Larsen, S. and Myrmel, M. 2008. Detection of norovirus genotype I.3b and II.4 in bioaccumulated blue mussels using different virus recovery methods. International Journal of Food Microbiology 127(1-2): 53–59.

Croci, L., Dubois, E., Cook, N., De Medici, D., Schultz, A.C., China, B., Rutjes, S.A., Hoorfar, J. and van der Poel, W.H.M. 2008. Current methods for extraction and concentration of enteric viruses from fresh fruit and vegetables: towards international standards. Food Analytical Methods 1(2): 73–84.

De Roda Husman, A.M., Lodder-Verschoor, F., van den Berg, H.H., Le Guyader, F.S., van Pelt, H., van der Poel, W.H. and Rutjes, S.A. 2007. Rapid virus detection procedure for moleculartracing of shellfish associated with disease outbreaks. Journal of Food Protection 70(4): 967–974.

Dubois, E., Agier, C., Traoré, O., Hennechart, C., Merle, G., Crucière, C. and Laveran, H. 2002. Modified concentration method for the detection of enteric viruses on fruits and vegetables by reverse transcriptase-polymerase chain reaction or cell culture. Journal of Food Protection 65(12): 1962–1969.

D'Souza, D.H. and Joshi, S.S. 2016. Foodborne viruses of human health concern. *In*: Paul M. Finglas e Fidel Toldrá (eds.). Encyclopedia of Food and Health. Benjamin Caballero, ISBN 978-0-12-384953-3

Elizaquível, P., Aznar, R. and Sánchez, G. 2014. Recent developments in the use of viability dyes and quantitative PCR in the food microbiology field. Journal of Applied Microbiology 116: 1–13.

Fukuda, S., Sasaki, Y., Kuwayama, M. and Miyazaki, K. 2007. Simultaneous detection and genogroup-screening test for norovirus genogroups I and II from fecal specimens in single tube by reverse transcription-loop-mediated isothermal amplification assay. Microbiol. Immunol. 51: 547–550.

Fumian, T.M., Leite, J.P., Marin, V.A. and Miagostovich, M.P. 2009. A rapid procedure for detecting noroviruses from cheese and fresh lettuce. Journal of Virological Methods 155(1): 39–43.

Greninger, A.L. et al. 2015. Rapid metagenomic identification of viral pathogens in clinical samples by real-time nanopore sequencing analysis. Genome Med. 7: 99.

Guévremont, E., Brassard, J., Houde, A., Simard, C. and Trottier, Y.L. 2006. Development of an extraction and concentration procedure and comparison of RT-PCR primer systems for the detection of hepatitis A virus and norovirus GII in green onions. Journal of Virological Methods 134(1-2): 130–135.

Hamza, I.A. and Bibby, K. 2019. Critical issues in application of molecular methods to environmental virology. J. Virol. Methods 266.

Haramoto, E., Katayama, H., Oguma, K. and Ohgaki, S. 2005. Application of cation-coated filter method to detection of noroviruses, enteroviruses, adenoviruses, and torque teno viruses in the Tamagawa River in Japan. Applied and Environmental Microbiology 71(5): 2403–2411.

Heid, C.A., Stevens, J., Livak, K.J. and Williams, P.M. 1996. Real time quantitative PCR. Genome Research 6(10): 986–994.

Hellmér, M., Paxéus, N., Magnius, L., Enache, L., Arnholm, B., Johansson, A. et al. 2014. Detection of pathogenic viruses in sewage provided early warnings of hepatitis A virus and norovirus outbreaks. Appl. Environ. Microbiol. 80: 6771–6781.

Jothikumar, N., Lowther, J.A., Henshilwood, K., Lees, D.N., Hill, V.R. and Vinjé, J. 2005. Rapid and sensitive detection of noroviruses by using TaqMan-based one-step reverse transcription-PCR assays and application to naturally contaminated shellfish samples. Applied and Environmental Microbiology 71(4): 1870–187.

Kurdziel, A.S., Wilkinson, N., Langton, S. and Cook, N. 2001. Survival of poliovirus on soft fruit and salad vegetables. Journal of Food Protection 64(5): 706–709.

Le Guyader, F.S., Parnaudeau, S., Schaeffer, J., Bosch, A., Loisy, F., Pommepuy, M. and Atmar, R.L. 2009. Detection and quantification of noroviruses in shellfish. Appl. Environ. Microbiol. 75: 618–624.

Li, J.W., Xin, Z.T., Wang, X.W., Zheng, J.L. and Chao, F.H. 2002. Mechanisms of inactivation of hepatitis A virus by chlorine. Applied and Environmental Microbiology 68(10): 4951–4955.

Logares, R. et al. 2012. Environmental microbiology through the lens of high-throughput DNA sequencing: Synopsis of current platforms and bioinformatics approaches. J. Microbiol. Methods 91: 106–113.

Love, D.C., Casteel, M.J., Meschke, J.S. and Sobsey, M.D. 2008. Methods for recovery of hepatitis A virus (HAV) and other viruses from processed foods and detection of HAV by nested RT-PCR and TaqMan RT-PCR. International Journal of Food Microbiology 126(1-2): 221–226.

Ma, J.F., Straub, T.M., Pepper, I.L. and Gerba, C.P. 1994. Cell culture and PCR determination of poliovirus inactivation by disinfectants. Applied and Environmental Microbiology 60(11): 4203–4206.

Mattison, K. and Bidawid, S. 2009. Analytical methods for food and environmental viruses. Food and Environmental Virology 1(3-4): 107–122.

McDaniel, L., Breitbart, M., Mobberley, J., Long, A., Haynes, M., Rohwer, F. and Paul, J.H. 2008. Metagenomic analysis of lysogeny in Tampa Bay: implications for prophage gene expression. PLOS ONE 3(9): e3263.

Morton, V., Jean, J., Farber, J. and Mattison, K. 2009. Detection of noroviruses in ready-to-eat foods by using carbohydrate-coated magnetic beads. Applied and Environmental Microbiology 75(13): 4641–4643.

Nieuwenhuijse, D.F. and Koopmans, M.P. 2017. Metagenomic sequencing for surveillance of food- and waterborne viral diseases. Front Microbiol. 8: 230.

Nocker, A., Cheung, C.Y. and Camper, A.K. 2006. Comparison of propidium monoazide with ethidium monoazide for differentiation of live vs. dead bacteria by selective removal of DNA from dead cells. Journal of Microbiological Methods 67: 310–20.

Nuanualsuwan, S. and Cliver, D.O. 2002. Pretreatment to avoid positive RT-PCR results with inactivated viruses. Journal of Virological Methods 104(2): 217–225.

Nuanualsuwan, S. and Cliver, D.O. 2003. Capsid functions of inactivated human picornaviruses and feline calicivirus. Applied and Environmental Microbiology 69(1): 350–357

Perelle, S., Cavellini, L., Burger, C., Blaise-Boisseau, S., Hennechart-Collette, C., Merle, G. and Fach, P. 2009. Use of a robotic RNA purification protocol based on the NucliSenseasyMAG for real-time RT-PCR detection of hepatitis A virus in bottled water. Journal of Virological Methods 157(1): 80–83.

Pexara, A. and Govaris, A. 2020. Foodborne viruses and innovative non-thermal food-processing technologies. Foods 9: 1520; doi:10.3390/foods9111520.

Pintó, R.M. and Bosch. 2008. Rethinking virus detection in food. pp. 171–188. Koopmans, M., Cliver, D.O. and Bosch, A. (eds.). Foodborne Viruses: Progress and Challenges. ASM Press, Washington, DC, USA

Rodríguez, R.A., Pepper, I.L. and Gerba, C.P. 2009. Application of PCR-based methods to assess the infectivity of enteric viruses in environmental samples. Applied and Environmental Microbiology 75(2): 297–307.

Rodríguez-Lázaro, D., Lombard, B., Smith, H.V., Rzeżutka, A., D'Agostino, M., Helmuth, R., Schroeter, A., Malorny, B., Miko, A., Guerra, B., Davison, J., Kobilinsky, A., Hernández, M., Bertheau, Y. and Cook, N. 2007. Trends in analytical methodology in food safety and quality: monitoring microorganisms and genetically modified organisms. Trends Food Sci. Technol. 18: 306–319.

Rodríguez-Lázaro, D., Ruggeri, F.M., Sellwood, J., Nasser, A., Sao Jose Nascimento, M., D'Agostino, M., Santos, R., Saiz, J.C., Rzeżutka, A., Bosch, A., Gironés, R., Carducci, A., Muscillo, M., Kovac, K., Diez-Valcarce, M., Vantarakis, A., Cook, N., von Bonsdorff, C.H., Hernández, M. and van der Poel, W.H.M. 2012. Virus hazards from food and the environment. FEMS Microbiol. Rev. 36: 786–814.

Rosario, K. and Breitbart, M. 2011. Exploring the viral world through metagenomics. Curr. Opin. Virol. 1: 1–9.

Roux, S., Tournayre, J., Mahul, A., Debroas, D. and Enault, F. 2014. Metavir 2: new tools for viral metagenome comparison and assembled virome analysis. BMC Bioinformatics 15: 76.

Rutjes, S.A., Italiaander, R., van den Berg, H.H., Lodder, W.J. and de Roda Husman, A.M. 2005. Isolation and detection of enterovirus RNA from large-volume water samples by using the

NucliSens miniMAG system and real-time nucleic acid sequence-based amplification. Applied and Environmental Microbiology 71(7): 3734–3740.

Schmittgen, T.D., Zakrajsek, B.A., Mills, A.G., Gorn, V., Singer, M.J. and Reed, M.W. 2000. Quantitative reverse transcription-polymerase chain reaction to study mRNA decay: comparison of endpoint and real-time methods. Anal. Biochem. 285: 194–204.

Severi, E., Verhoef, L., Thornton, L., Guzman-Herrador, B.R., Faber, M., Sundqvist, L. et al. 2015. Large and prolonged food-borne multistate hepatitis A outbreak in Europe associated with consumption offrozen berries, 2013 to 2014. Euro Surveill. 20: 21192.

Simonet, J. and Gantzer, C. 2006a. Degradation of the Poliovirus 1 genome by chlorine dioxide. Journal of Applied Microbiology 100(4): 862–870.

Simonet, J. and Gantzer, C. 2006b. Inactivation of poliovirus 1 and F-specific RNA phages and degradation of their genomes by UV irradiation at 254 nanometers. Applied and Environmental Microbiology 72(12): 7671–7677.

Stals, A., Baert, L., De Keuckelaere, A., Van Coillie, E. and Uyttendaele, M. 2011b. Evaluation of a norovirus detection methodology for ready-to-eat foods. International Journal of Food Microbiology 145(2-3): 420–425.

Stals, A., Baert, L., Van Coillie, E. and Uyttendaele, M. 2012. Extraction of foodborne viruses from food samples: a review. International Journal of Food Microbiology 153(1-2): 1–9.

Suttle, C.A. 2007. Marine viruses—major players in the global ecosystem. Nat. Rev. Microbiol. 5: 801–12.

Temmam, S., Davoust, B., Chaber, A.-L., Lignereux, Y., Michelle, C., Monteil-Bouchard, S. et al. 2017. Screening for viral pathogens in African simian bushmeat seized at a French airport. Transb. Emerg. Dis. 64: 1159–1167.

Tian, P., Engelbrektson, A. and Mandrell, R. 2008. Two-log increase in sensitivity for detection of norovirus in complex samples by concentration with porcine gastric mucin conjugated to magnetic beads. Applied and Environmental Microbiology 74(14): 4271–4276.

Venkatesan, B.M. and Bashir, R. 2011. Nanopore sensors for nucleic acid analysis. Nat. Nanotechnol. 6: 615–624.

Wang, Y., Zhu, N., Li, Y., Lu, R., Wang, H., Liu, G., Zou, X., Xie, Z. and Tan, W. 2016. Metagenomic analysis of viral genetic diversity in respiratory samples from children with severe acute respiratory infection in China. Clin. Microbiol. Infect. 22: 458 e1-9.

Wylie, K.M., Weinstock, G.M. and Storch, G.A. 2012. Emerging view of the human virome. Translational Research 160: 283–290.

Wylie, K.M., Weinstock, G.M. and Storch, G.A. 2013. Virome genomics: a tool for defining the human virome. Curr. Opin. Microbiol. 16: 479–84.

Yoneyama, T., Kiyohara, T., Shimasaki, N., Kobayashi, G., Ota, Y., Notomi, T., Totsuka, A. and Wakita, T. 2007. Rapid and real-time detection of hepatitis A virus by reverse transcription loop-mediated isothermal amplification assay. J. Virol. Methods 145: 162–168.

Chapter 6

Concentration and Detection of Enteric Viruses in Aqueous Samples

Kareem R. Badr,[1] *Rachel Siqueira de Queiroz Simões,*[2,3,4]
David Rodríguez-Lázaro[3,4,*] and *Elmahdy M. Elmahdy*[1]

1. Introduction

For at least 2500 year b.c., the human species has been in the habit of seeking quality water through the digging of artisan wells (Baker and Michael 1981, Diniz-Mendes 2018). In Greece, Hippocrates had already observed the relationship between water quality and the development of diseases. The implementation of treatments for disinfection in water, occurred for the first time with the discovery of the British scientist John Snow and William Budd through classic epidemiological techniques. In the aquatic environment, each cubic millimeter of water in the ocean contains several million virus particles. They are responsible for the biomass and ecological regulation of water ecosystems and estimates suggest there are about 10^{30} virions across the ocean. On the other hand, not all marine viruses are associated with infections in humans. The main concern is that the viruses present in the gastrointestinal tract of infected individuals can be excreted in the feces in large quantities and thus are able to contaminate water sources for human consumption or recreational activities.

[1] Environmental Virology Laboratory, Water Pollution Research Department, Environmental and Climate Change Research Institute, National Research Centre, Dokki, 12622, Giza, Egypt.
[2] Institute of Technology in Immunobiologicals, Bio-Manguinhos, Oswaldo Cruz Foundation, Fiocruz, Avenida Brasil, 4365, Manguinhos, Rio de Janeiro 21040-900, Brazil.
[3] Microbiology Division, Faculty of Sciences, University of Burgos, 09001 Burgos, Spain.
[4] Research Centre for Emerging Pathogens and Global Health, University of Burgos, 09001 Burgos, Spain.
* Corresponding author: drlazaro@ubu.es

Enteric viruses are inert particles outside of their hosts and the transmission from one host to another depends on the degree of their resistance. They remain infectious under various environmental conditions that may affect their persistence. Some of the primary factors affecting the survival of viruses in aqueous matrices are temperature, ionic strength, chemical constituents, microbial antagonism, the sorption status of the virus, and the type of virus. In 1909, poliovirus was identified as the causative agent of acute paralysis that affected people's health all over the world (Landsteiner and Popper 1909). As the polio epidemic continued to be a scourge in the world, scientists focused on determining how the disease was transmitted. In the early 1940s, it was thought that poliovirus could only be transmitted via the person-to-person route, after an outbreak of gastrointestinal illness associated with wading and swimming in a local creek in the USA. This was the first demonstration of viral transmission via the water route (Toomey 1945). An American researcher Joseph L. Melnick detected the presence of poliovirus in rhesus monkeys after experimental inoculation with sewage samples demonstrating the correlation of viral particles present in sewage. Another outbreak of the hepatitis E virus was associated also with water transmission routes in India, resulting in nearly 30,000 infections and 73 deaths (Dhopeshwarkar et al. 1957, Naidu and Viswanathan 1957, Viswanathan and Sidhu 1957). In the 1950s, polioviruses, coxsackieviruses and echoviruses were also detected and isolated from sewage and contaminated water samples. In 1965, the first International Conference on Transmission of Viruses by Water was performed in the United States, and in the same year the presence of poliovirus in Paris city water systems was described at the second International Conference on Water (Chang et al. 1968). From this historic milestone, other conferences were carried out such as "Viruses in water" in Mexico in 1974 and in the "Enteric viruses in water" in Israel in 1982, where methods for detecting low concentrations of viruses in aquatic environments were addressed. In 1979, the World Health Organization issued the first publication on methods for virus isolation from water samples (WHO 1979) followed by another publication that included methods for virus concentration and detection from water samples (Greenberg and Taras 1981). So, the development of techniques in the area of environmental microbiology gained recognition and notoriety (Baker and Michael 1981, Diniz-Mendes 2018).

Various studies were conducted aiming at these findings and it became clear that viruses could be transmitted by water and affect human health. Most of the enteric viruses such as human adenoviruses, caliciviruses (norovirus and sapovirus), enteroviruses, rotaviruses, the hepatitis A and E viruses, and others can be spread by the fecal-oral route (Kern et al. 2013). According to WHO, several enteric pathogenic viruses such as Adenovirus, Hepatitis A virus (HAV), Hepatitis E virus (HEV) and Rotavirus were detected in the pan-European region, and were associated with drinking-water and pool-water during recreational activities. For example, sapoviruses and enteroviruses were detected in an outbreak in Finland and human noroviruses (GI/GII), rotaviruses, sapoviruses, human astrovirus and adenovirus were detected after a gastrointestinal outbreak at Denmark using microbiological analysis of water. Scientists continue to develop the best approach for monitoring the virus occurrence in the environment, especially in water, virus fate, and transport in a different ecosystem, and to quantify their associated risk to humans in water. Some viruses are more resistant to water and sewage treatment processes, which makes the monitoring of recreational water essential for surveillance of quality standards and control the environmental impact of these waterborne diseases on public health. To assess the presence of viruses in water, several studies tried to show the best way to detect viruses in water, from the concentration of samples to virus identification using different detection methods. While several technologies have been developed and refined, it has proven difficult to achieve detection of all relevant virus types across the range of water matrices. This chapter describes the concentration methods of viruses from water, including the primary concentration methods that have been developed along with different studies that aimed to enhance the detection of enteric viruses in water, in addition to describing the most common detection methods using modern molecular biology techniques.

2. Concentration and recovery of viruses from water

2.1 Early concentration techniques

The early studies aimed to determine the size of virion particles and provide the techniques that lead to the isolation of the virus from the samples. One of the first studies was conducted by Elford that investigated the filtration of viruses based on viral adsorption to collodion membranes (Elford 1931). In 1952, studies used ion-exchange resins to concentrate poliovirus from environmental water and sewage samples (Kelly 1953). Flocculation procedures have been also used to remove viruses from river water based on the formation of a metalliccation-protein complex and its aggregation, which then led to the use of second-step concentration techniques from large-volume water sampling (Chang et al. 1958, Chang 1968). Several studies used volunteers who drank water containing infective fecal material clarified or not by alum flocculation to assess the concentration of the virus ingested and to evaluate the efficiency of the flocculation process (Chang 1968). At the beginning of the 1960s, there was a focus on the development of new methods to isolate viruses from the environment. Different studies described concentration methods based on the volume of water samples, from small to large quantities. Most of these studies were designed to recover viruses from small water volumes and grab sample collection. This was one of the early methods for assessing viral presence in water, which later proved to be ideal for wastewater with high viral loads (Kelly 1953, Melnick et al. 1954, Gravelle and Chin 1961). In this context, various methods have been used such as aqueous polymer two-phase separation, hydroextraction, soluble alginate ultrafilter membranes, and ultracentrifugation (Table 1).

Aqueous polymer two-phase separation employed liquid-liquid partitioning to stimulate the movement of viruses from water samples to an organic solution like dextran and polyethylene glycol (PEG). By this time, various studies reported promising results using this method (Shuval et al. 1967, 1969), although there were some drawbacks such as the use of minimum sample size (1 L) or inhibition of some viruses like echovirus and Coxsackievirus serotypes (Grinrod and Cliver

Table 1. Methods for virus concentration in water samples.

Techniques	Sample type	Initial volume	Virus	Recovery rate	Advantages	Disadvantages	References
Aqueous polymer two-phase separation	All waters	1–1000 mL	Echovirus, Coxsackievirus	Medium	Low multiplicities of virus detected	Treats small volumes only; Inhibition of some virus's activity.	Shuval et al. (1967), Shuval et al. (1969), Grinrod and Cliver (1970), Grinstein et al. (1970)
Hydroextraction	All waters	1–1000 mL	Echovirus, Coxsackievirus	Medium	Moderate to high virus recovery	Treats small volumes only; Concentrate also the toxic substances present in the water.	Shuval et al. (1967)
Soluble alginate ultrafilter membranes	All waters	1–1000 mL	Poliovirus	Medium	Non-cytotoxic; No virus inactivation	Treats small volumes only; Turbid waters should be filtered first; viruses associated with solid compounds could not be concentrated	Gartner (1967) Nupen (1970)
Ultracentrifugation	All waters	1–1000 mL	feline calicivirus, hepatitis A virus,	Medium	Fractionation of diverse virus types possible	Treats small volumes only; Exorbitant cost; Require large-scale instrument Excessive processing times	Cliver and Yeatman (1965), Hill et al. (1971)
Electronegative	Wastewater Fresh water	1–5 L 1–20 L	Coxsackievirus, poliovirus, reovirus, echovirus, norovirus	Medium to high	Economical; Can filter large volumes even in more turbid waters before clogging occurs; Has been tested with an array of enteric viruses; High recoveries for commonly tested enteroviruses	Requires preconditioning of water sample prior to filtration; highly sensitive to organic matters in aqueous samples	Berg et al. (1971) Haramoto et al. (2009)

Category	Filter	Water type	Volume	Viral pathogens	Recovery	Advantages	Limitations	References
Electropositive	1MDS	Fresh water Tap Water	1–20 L	Poliovirus bacteriophage	Medium to high	No preconditioning of water sample required; Has been tested with an array of enteric viruses; Can filter large volumes even in more turbid waters before clogging occurs	Exorbitant cost per filter; Highly sensitive to organic matters in aqueous samples	Sobsey and Glass (1980) Dahling (2002) Cashdollar and Dahling (2006)
	NanoCeram	Deionized water Fresh water	100 L	Poliovirus, Coxsackievirus, Echovirus, Bacteriophage,	Medium to high	Economical; Comparable recoveries to 1 MDS for viruses tested; No preconditioning of water sample required	Clogging problems in more turbid waters; Limited data available with multiple viral pathogens	Karim et al. (2009) Gibbons et al. (2010) Bennett et al. (2010) Ikner et al. (2011)
	ViroCap	Fresh water Wastewater	1–20 L	Poliovirus, Coxsackievirus, Echovirus	Medium to high	Economical; No preconditioning of water sample required; Easy to use; Field-deployable	Clogging problems with turbid water; Limited volumes due to filter's size; Limited data available with multiple viral pathogens	Karim et al. (2009) Fagnant et al. (2017)
	Glass Wool	Fresh water Tap Water	5–20 L	Poliovirus, Hepatitis A virus, transmissible gastroenteritis virus, Adenovirus	Medium to high	Economical; No preconditioning of water sample required; Easy to use; Field-deployable	Laboratory data variation depend on the filter apparatus that will be used; Clogging problems with turbid water; Limited data available with multiple viral pathogens	Vilagines et al. (1997) Blanco et al., (2019) Lambertini et al. (2008)
Ultrafiltration - Hollow-Fibre		Wastewater groundwater Fresh water	1–1000 mL 10–50 L	echovirus poliovirus bacteriophage	Medium to high	Multi-pathogen concentration; Economical; No preconditioning of water sample required	Clogging problems with turbid water; Difficult to deploy in the field; The filtration rate is usually slow; Limited data available with multiple viral pathogens	Olszewski et al. (2005) Hill et al. in (2009) Holowecky et al. (2009) Gibson and Schwab (2011)

1970). Hydroextraction is a method that requires the transfer of the water sample into a dialyzing bag. Its volume is subsequently reduced by a hydrophilic agent surrounding the bag (usually PEG) that extracts water from the sample using direct virus precipitation with a recovery rate of 40–50% (Shuval et al. 1967, Van Alphen et al. 2014). Some studies used hydroextraction as a secondary-step concentration method after initial adsorption of simian rotavirus SA1 from dechlorinated 10I tap water which, consequently, improved the recovery detection rate (Pádraig Strappe 1991). The main disadvantage of using this method is that the toxic substances present in water can also concentrate in the final extract volume and, consequently, will affect cell culture toxicity.

Another method based on soluble ultrafilters sodium alginate was able to concentrate poliovirus from various water sources (Gartner 1967, Nupen 1970). In this method, the filters were subsequently dissolved in sodium citrate to elute the virus and then applied to cell culture infectivity assays for detection. In this case, the elution step is determined based on the primary method since the development of the soluble sodium alginate filters as the primary concentration method should be followed by sodium citrate as the eluting reagent. Additionally, because of the nature of soluble alginate filter materials, water samples with high turbidities, such as wastewater, must be pretreated by passing through a series of filters to avoid clogging during the concentration process.

Ultracentrifugation was also used for the same purpose by Cliver and Yeatman in 1965, but the method had little practical value due to the elevated cost of equipment, the ability to process only small water volumes, and excessive processing times (Hill et al. 1971). This method exhibits a similar and sometimes greater recovery of viruses from environmental samples than other methods when used as a secondary concentration step. Schultz et al. (2011) had a better recovery rate for feline calicivirus and hepatitis A virus with ultrafiltration method as a secondary concentration step after following primary concentration by a positive charge filter. Several advantages of using ultracentrifugation as a concentration method to detect enteric viruses. There is no required pretreatment step without any precise or new materials, containers may be reused, so it is considered a low-cost method.

Various studies have shown that using these methods was not feasible to concentrate the virus from large quantities of water (Grabow 1968, Rao and Labzoffsky 1969). Therefore, new methods were used in large water samples based on adsorption properties of some materials, such as; polyelectrolytes that were investigated for their capacity to remove viruses from distilled water (Johnson et al. 1967), sewage, and streams receiving sewage effluents (Grinstein et al. 1970); precipitable salts that showed the ability to concentrate several enterovirus types including poliovirus 1 and echovirus 7 (Wallis and Melnick 1967); in addition to the use of cotton gauze fibers that were used to concentrate poliovirus from large volumes of water during an epidemic in Paris, France (Coin 1967).

2.2 Commonly used concentration techniques

Viral particles can be concentrated based on the same general principles like difference in electrical charge and molecular mass. Thus, the main concentration methods involve (i) adsorption and elution to electrically charged solid matrices, (ii) precipitation and (iii) ultrafiltration (Diniz-Mendes 2018). The most common methods nowadays are mainly based on primary followed by secondary concentration of water samples. Adsorption/elution and ultrafiltration methods are considered the two main methods used in these techniques. The adsorption/elution method generally uses one of both filter types; electronegative (EN) or electropositive (EP) filters (Cashdollar and Wymer 2013, Ikner et al. 2012) in which the water sample passes through the filter, while the virus particles bind to the surface of the filter due to electrostatic forces. However, the ultrafiltration method has been used for the concentration of a wide range of water samples for the detection of various pathogens (Table 1).

2.2.1 Adsorption/elution methods

In this method, as described by Wallis and Melnick (1967), the viral particles are adsorbed on filter media through charge interaction, followed by subsequent elution of these viruses by a pH-adjusted solution. So, this method was widely developed until it had a greater potential to process large volumes of water (Anderson et al. 1967, Stevenson 1967). This method can concentrate viral particles from large water volumes, even thousand litres or more, and can be applied to water supply systems or water springs (Diniz-Mendes 2018), and also to small volumes of approximately 0.1 to 5 litres, to assess samples from rivers or sewage. In these procedures, viral particles adhere to a solid matrix due to electrical interactions by adjusting pH conditions and ionic strengths that are different for each type of sample (Diniz-Mendes 2018). Many considerations can directly or indirectly affect the use of this method, such as the water matrix composition, type of filter used to adsorb the virus in addition to the elution solutions. Considering the water matrix, these methods are used mainly for water samples unlike sewage samples, which can clog filters. As for the elution solution, the most common elution solution typically used in this method contains beef extract. However, solutions containing amino acids and/or salts have also been used to remove adsorbed virus from the filter (Farrah and Bitton 1979, Chang et al. 1981). Since viruses in water normally have a net negative surface charge, depending on the type of filter utilized, either the filters or the water sample must be conditioned before filtration to allow adsorption.

There are two factors that can facilitate adsorption: the pH, that must be around 3 and 5, except in the case of glass wool, when the adsorption occurs at neutral pH, and the presence of polyvalent cations such as magnesium and aluminium that can elevate the pH, leading to changes in the physicochemical conditions and, consequently, in the interactions between the viruses and the matrix. The particles are eluted in small volumes of alkaline solutions such as meat extract, skim milk, aluminium sulphate or sodium hydroxide with a pH above 10 (Diniz-Mendes 2018). The selection of the correct filter can also be considered a key factor, and its selection depends mainly on the type and composition of the virus. The virus usually consists of an outer shield called the capsid, which surrounds the genomic materials of the virus forming various shapes. Enteric viruses have a significant variation in the number of proteins present in their capsid, which then influences the size and charge of the proteins that make up the capsid. Enterovirus is considered the smallest enteric virus in diameter (30 nm) while adenovirus is about 100 nm in size. The isoelectric point's properties also vary from 28 to 80 for hepatitis A and rotavirus respectively (Michen and Graule 2010). There are two basic filter types used to adsorb viruses: electronegative filters and electropositive filters.

Electronegative filters

Cellulose nitrate HA membranes (Millipore) were reported as the first membrane filters used to recover influenza viruses from water (Metcalf 1961). Coating the cellulose nitrate membranes with proteinaceous or gelatin-based substances can enhance virus passage through the membrane (Cliver 1965). Different studies used to utilize this type of membrane (0.45-μm pore size) to concentrate enteric viruses after adding low molarity concentrations of $NaCl_2$ and magnesium chloride ($MgCl_2$) salts to water sample at pH 5 (Wallis and Melnick 1967, Ver et al. 1968), or reduced pH to 3.5 by adding 1 N HCl (Kern et al. 2013), which also enhanced the adsorption capacity of other viruses such as poliovirus 1 (Wallis and Melnick 1967, Rao and Labzoffsky 1969). In addition, water samples can be concentrated using glass wool (type 725) and mixed membrane of 142 mm diameter with a glass wool pre-filter, diameter 125 mm (Kern et al. 2013). Coxsackievirus B3, poliovirus 1, reovirus 1, and echovirus 7 were concentrated from raw waters using McIlvaine's Buffer as described by Berg et al. (1971). In 2009, a study developed a method for using an electronegative filter in conjunction with aluminium or magnesium to recover norovirus using different water sources (Haramoto et al. 2009). In this study, norovirus was recovered from MilliQ water (186%), tap water (80%), bottled water (167%), river water (15%), and pond water (39%) effectively. While Victoria et al. (2009) used negatively charged membranes to recover noroviruses and human astroviruses from river water (64%), mineral water (18%), seawater (14%), and tap water (3%).

The use of electronegative filters as a concentration method in water samples shows relatively high recoveries of waterborne viruses as well as they are cost-effective. Still, the major disadvantage of using this type of filter is the difficulty of using large-volume water samples. In addition, the adjustment of pH for the large volume may be difficult to apply at the field and lead to limits in the volume of sample to do this adjustment at the laboratory, since most laboratories are not equipped to handle hundreds of litres of water. Before sample collection, electronegative filters, rather than water samples, can be preconditioned. But the main disadvantage is the possibility for higher variability in filter surface charge, which introduces another variable when assessing filter performance and makes frequent monitoring difficult unless strong quality assurance mechanisms are in place.

Electropositive filters

Precipitation occurs in the presence of positively charged coagulants such as aluminium hydroxide, forming a complex with the negatively charged viral particles. Whereas electronegative filters rely on the manipulation of the water sample to cause a net positive surface charge of a viral particle, electropositive filters work by relying on the innate negative charge of the viral particle in water. Otherwise, it does not typically require preconditioning of a water sample before filtration and can filter large volumes of water (> 1000 l) at high filtration rates without clogging in most cases. Cartridge/type filter, 1 MDS filter, was developed and has been popular as an electropositive filter. Early studies on poliovirus (PV) using this type of filter showed a recovery efficiency similar to that of the electronegative filter in spiked tap water samples (Sobsey and Glass 1980). The recovery efficiency varied from 36% (Karim et al. 2009) to 95% with seeded PV in river water (Dahling 2002), and 67% of tap water (Karim et al. 2009). The major disadvantage to the 1 MDS filter is the cost, so study showed that it's possible to wash and reuse the 1 MDS filter after washing 1–3 times with average recoveries ranging from 30 to 38% for tap water samples (Cashdollar and Dahling 2006, Polaczyk et al. 2007), and for those filters used 1–2 times, average recoveries ranged from 53 to 72% for river water samples (Cashdollar and Dahling 2006).

Another electropositive filter that has recently entered the market is the NanoCeram filter. This filter is offered in cartridge format or as a self-contained cartridge under the name Virocap, and is much more economical. The recovery ability of this filter reached 84% for PV (Karim et al. 2009), and with a developed protocol for the elution step, it reaches 54% for PV, 27% for coxsackievirus B5, and 32% for echovirus 7 when seeded in tap water samples. Gibbons et al. (2010) reported recoveries of > 96% for male-specific coliphages and noroviruses but < 3% for adenoviruses when using the NanoCeram filter to concentrate from seawater. The deionized water showed recoveries of 65% and 63% from seawater for bacteriophage MS2 (Bennett et al. 2010) and 37% from deionized water and 44% from seawater for PV when using the Virocap filter. Another study demonstrated recoveries of 45–56% for bacteriophage MS2, 66% for poliovirus 1, 84% for echovirus 1, 77% for coxsackievirus B5 and 14% for adenovirus 2 from 20 l spiked tap water samples when using the NanoCeram filter (Ikner et al. 2011).

The glass wool filter system became a promising tool for concentrating viruses from large volumes of water at the beginning of 1990 with an easy way to apply, just the glass wool is packed into a column, and then the column is connected to the water source. The recovery efficiency of this filter ranges from 72 and 75% when applied to recover poliovirus from drinking water and seawater, respectively (Vilagines et al. 1997). Another study showed high recoveries for poliovirus (98%), adenovirus 41 (28%), and norovirus GII (30%) from tap water (Lambertini et al. 2008). Like NanoCream filters, glass wool is easy to use, inexpensive and does not need any preconditioning of the water sample before filtration as well as is a perfect option for large-volume sampling and/or in-line continuous sampling. Another study showed that the recovery rate of Hepatitis A virus (HAV) and transmissible gastroenteritis virus (TGEV) could be developed by modifying the elution condition such as pH, contact timing and adding detergent (Blanco et al. 2019).

2.2.2 Ultrafiltration

The alternative technique to adsorption–elution methods is the ultrafiltration method which is based on size exclusion (among 30 to 100 kDa) rather than electrostatic interactions between electronegative viruses and negatively or positively charged filter surfaces. The main advantage to this method is that the sample does not require any preconditioning procedures before filtering, but at the same time it requires costlier equipment and is generally not feasible for field sampling in addition to its tendency to clog under conditions when particulates exceed 16 g/L concentration (Olszewski et al. 2005). The most common matrices are filters, membranes, capillaries, and hollow fibers. Filtration can occur by pressing samples through a filter or by dialysis. Some organic solutions (as beef extract) and sodium polyphosphates could be used as pretreatment solutions to prevent the viral particle to be adsorbed to the ultrafilter membranes due to their hydrophobic behaviour (Berman et al. 1980, Hill et al. 2005).

Ultrafiltration can be a dead-end (single pass of the sample through the ultrafilter) or tangential flow (multiple circulations of the sample through the ultrafilter). Typically, the ultrafilter is then backwashed to remove any micro-organisms that are retained on the filter, and this backwash is combined with the retentate of the final sample for analysis. A hollow-fiber ultrafiltration system is an ideal method to be used for virus concentration from large volumes of water. The method is conducted by passing the water sample through capillaries or hollow fibers or through flat sheets using tangential flow, all of which have nominal molecular weight cut-offs of 30–100 kDa. Because of the pore size, water and low molecular weight substances are allowed to pass through the fibers and into the filtrate, whereas larger substances, such as viruses and other microorganisms, are trapped and retained in the retentate.

Using this method, Olszewski and co-workers were able to get a recovery of 57–71% for bacteriophage T1, 70–74% for bacteriophage PP7, and 82–95% for poliovirus 2 from groundwater samples (Olszewski et al. 2005). While Hill et al. (2009) mentioned recoveries for echovirus 1 (58%), bacteriophage MS2 (100%), and bacteriophage phi X174 (77%) from seeded tap water samples. In another study, the recovery of bacteriophages MS2 ranged 52–88% and for phi-X174 was 55–95% depending on the type of filter used (Holowecky et al. 2009). Gibson and Schwab (2011) using surface water as a matrix showed recoveries of 48 and 84% for bacteriophage MS2, 43 and 80% for bacteriophage PRD1, and 40 and 16% of poliovirus under high and low seed levels, respectively. In the same study, the recovery efficiency for Murine norovirus type 1 in surface water samples was 74% using low seed levels, and 42% in drinking water samples using high speed levels.

3. Methods for detection of viruses in water

The ability of viruses to remain infectious in aquatic environments is normally high, even in extremely low concentrations, and can survive in water for several months depending on the virus type (Wyn-Jones and Sellwood 2001). Quantifying the number of infectious viruses in water and wastewater is necessary to determine the risks associated with exposure and also the degree of treatment needed to reduce these risks to an acceptable level (Gall et al. 2015, Gerba et al. 2017). Infectious viruses are defined as those capable of replicating in cell culture and thus, have the potential to replicate in humans and animals and cause disease. The detection and quantification methods for these viruses are based on infectivity-based techniques where the virus undergoes at least partial multiplication in cell culture, or by one of the molecular biology techniques. As is known, most of these viruses are not adapted to cultivation in the cell culture, therefore, modern molecular biology techniques such as PCR and qPCR are considered the most promising techniques in viral monitoring.

Detection of virus infectivity as well as knowledge of the number of infectious viruses in water is traditionally done by inoculating cell cultures with the concentrated sample and allowing the virus to multiply in the cells so that they are killed. Plaque assay and tissue culture infective dose assay are usually used to describe and determine the effect of the inoculated virus on the cells. In the plaque assay method, the cytopathic effect (CPE) of many waterborne viruses and some other types could be visible by the naked eye. If the cell culture is inoculated with the sample under the liquid assay method it should be possible to detect some enteroviruses such as poliovirus, Coxsackie B, echoviruses, in addition to some adenoviruses and reoviruses. Other viruses need special treatment such as HAV that may be detected using this technique only after prolonged incubation of cultures. However, it is not an applicable approach to detect HAV in a water sample. Various approaches have been developed to assess the infectivity of waterborne enteric viruses using molecular methods, but they are specific to the virus and the mechanism of virus inactivation (Leifels et al. 2016, Randazzo et al. 2018, Leifels et al. 2019).

In addition, the observation of the cytopathic effect on infected cells, the visualization of viral particles by electron microscopy, hemagglutination and hemadsorption assays, direct and indirect immunofluorescence, viral neutralization assays, enzyme immunoassays and molecular tests are also methods of verifying the presence of waterborne viruses. This last method is the most important tool for detecting enteric viruses in water and monitoring of viral particles in concentrated samples, recovered from the environment. The modern molecular biology techniques which are commonly used these days to detect the genomic materials of a target virus are mainly based on qualitative methods by traditional PCR techniques or quantitative methods whether they are semi-quantitative, such as end-point dilution assays or fully quantitative using real-time PCR. Over the past few years, the analysis of samples detected by conventional PCR and confirmed by Southern Blot, nested or semi-nested PCR, real-time PCR, genome sequencing has shown the wide diversity of this enteric viruses' extensive spread in several aquatic ecosystems.

3.1 Cell culture techniques

Cell culture techniques involve the isolation and identification of the virus by propagation in permissive cell culture of animal or human origin, and are still the common approved method used to assess viral infectivity. However, some enteric viruses such as the noroviruses, cannot be cultured from the environmental samples and others, such as rotaviruses and hepatitis A virus are difficult to cultivate in cell culture systems. So, cell culture monolayer plaque assays are not applicable to most waterborne viruses. Buffalo green monkey (BGM) cell line is the most common line for enumeration of water-associated enteroviruses (EVs) (Dahling et al. 1974) and showed higher plaque assay titers for poliovirus, Coxsackie viruses B, some echovirus, and reoviruses (Morris 1985). A549 cells, derived from human lung tissue, support the growth of some adenoviruses (AdV) and have been used to study their effect on the cell proteome (Badr et al. 2019) or study virus infectivity using the integrated cell culture-PCR technique for rapid detection of infectious adenoviruses in water (Greening et al. 2002). A large survey was conducted by Patel and his group to study the susceptibility of some cell lines on EVs, where they found that HT-29 and SKCO-1 cell lines had a markedly wider sensitivity for EVs than primary monkey kidney (PMK) or RD cell cultures (Patel et al. 1985). Another study also used CACO-2 cells, along with RD, BGM, and human epithelial type 2 (HEp-2) cells to detect the infectious reoviruses, EVs, and AdV in a range of water types (Sedmak et al. 2005). Plaque assay and tissue culture infective dose methods are the commonly used techniques for enumerating waterborne infectious viruses and measuring their infectivity.

3.1.1 Plaque assay method

During virus study, practically there must be a step where it is necessary to measure the concentration of viruses in a sample. Plaque assay is considered the widely used approach for determining the number of infectious viruses. This technique was first developed to calculate the titers of bacteriophage stocks which was then modified by Renato Dulbecco in 1952 for use in animal virology, and it has since been used for determining the titers of other different viruses (Dulbecco 1952). In this method, plaques develop after a period of incubation of the infected monolayer cell and can be visible to the naked eye after dyeing the cells with a stain. The plaques are counted and the titer of a virus stock can be calculated in plaque-forming units (PFU) per millilitre. Differences between the percentage of virus concentration of PFU/litre in cold season and warm months have been reported in the same sample sites, even with an increase of over 10 times in warm periods (Chang et al. 1968). It should be noted that many variables such as host cell type, culture media type, Ion Concentration, Or Ph Might Affect The Apparent Concentration Of Virus Aliquot. For example, a virus stock may contain a titer of 10^5 PFU/mL on one type of cells, and the same stock might have a titer of 10^3 on another cell type. In addition, only viruses that cause visible damage to cells can be assayed in this way. The incubation period may be varied from one virus to another since the multiplicity of viruses vary. The plaques are continued after the first appearance and usually, it takes 2 to 5 days (SCA 1995) or 12 days until no new plaques appear as recommended by the United States Environmental Protection Agency (USEPA 2007). Using plaque assay technique to characterize human coronavirus NL63 (hCoV-NL63) infectivity, Herzog and his team showed that human colon carcinoma cells (CaCo-2) replicated the virus over 100 fold more efficiently than commonly used African green monkey kidney cells (LLC-MK2) (Herzog et al. 2008).

3.1.2 Tissue Culture Infective Dose (TCID$_{50}$)

As mentioned before, the plaque assay method cannot be applied for viruses that do not form plaques. So, there are several alternative methods available, including the tissue culture infective dose which is considered one of the end-point dilution assays that were used to measure virus titer before the development of the plaque assay. In this method, serial dilutions of a virus stock are prepared and inoculated onto replicate cell cultures using multi-well plates. After incubation, the number of cell cultures that are infected is determined for each virus dilution, usually by looking for the CPE. Normally, at high dilutions, none of the cell cultures will be infected because no virus particle would be present, while at low dilutions every cell culture may be infected. The desired dilution that will be considered at this moment is the dilution of the virus at which half of the cell cultures are infected and present CPE. Therefore, the Virus titer is calculated as the logarithm of the dilution of the virus-producing a CPE in 50% of the cultures, and the virus titer is expressed as 50% infectious dose (ID_{50}) per millilitre as described by Reed and Muench (1938). In animal studies, end-point dilution methods can also be used to determine the virulence of a virus in animals. The same methods are used but the serial dilutions of the virus are made and inoculated into multiple test animals instead of using cell culture. The assessment of the animal Infection can be determined by death or any other clinical symptoms such as fever, weight loss, or paralysis. The same concept is used to express the final results such as 50% lethal dose (LD_{50}) per ml or 50% paralytic dose (PD_{50}) per ml when lethality or paralysis are used as endpoints.

3.2 Molecular biology techniques

The polymerase chain reaction (PCR) technique is considered one of the main molecular biological detection techniques in the virology field, specifically for detection of enteric viruses in water samples. It is particularly useful in the detection of viral pathogens that do not multiply in cell culture such as most gastroenteritis viruses. The technique has been used to detect various enteric viruses including adenoviruses (Puig et al. 1994), HAVs (Graff et al. 1993), astroviruses (Marx et al. 1995), and rotaviruses (Gajardo et al. 1995). Magnetic silica bead kit or NucliSens® miniMAG™ system has been used to extract viral nucleic acids from concentrated samples because silica suspension reduces the risk of cross-contamination. Then, the samples are stored until further molecular assays (Kern et al. 2013, Wyn-Jones et al. 2011). Another study aimed to develop a quantitative approach using a group of RNA viruses including noroviruses, rotaviruses, enteroviruses, and reoviruses (Lodder and de Roda Husman 2005). They analyzed 10-fold serial dilutions of the extracted RNA and found that noroviruses were detected in 4–4900 PDU/litre (PCR-detectable units per litre) of river water and the higher titers were found in sewage. Moreover, there are some modifications of the conventional PCR method applied to enteric viruses: nested PCR for detection of human adenoviruses (HAdV) or RNA-dependent RNA polymerase (RdRp) by RT-PCR followed by a semi-nested PCR for detection of noroviruses with subsequent confirmation by sequencing.

Real-time PCR offers many advantages over cell culture-based techniques such as cost-effectiveness, time-efficiency, and the capacity to detect viruses that have slow- or not-growing in cell culture. In 2004, Laverick and co-workers designed a RT-quantitative PCR (RT-qPCR) to detect noroviruses in sewage, marine, and riverine recreational waters (Laverick et al. 2004). The virus was found in high amounts in sewage followed by effluent samples and sequentially in marine bathing water and recreational river waters. TaqMan probes are widely used for the detection of a variety of environmental samples such as Adenoviruses 40 and 41 (Jothikumar et al. 2005, Jothikumar et al. 2009, La Rosa et al. 2010). Using molecular methods, high prevalence of human adenoviruses (3.91×10^4 genome copies per litre) was detected in surface water samples. Other waterborne viruses like noroviruses and enteroviruses have been also detected by reverse transcription-PCR assays (Kern et al. 2013). Some enteric viruses can be detected by qPCR such as Adenovirus (adenovirus fiber gene in AdV40 and AdV41) with low detection limits-around 5–8 copies of AdV40/41 in wastewater, drinking water, recreational waters, and rivers (Jothikumar et al. 2005)—as well as others viruses such as Aichi virus, astrovirus, enterovirus, human norovirus, rotavirus, sapovirus, and hepatitis A and E viruses (with similar detection limits). Multiplex real time PCR is a method applied for simultaneous detection of enteric viruses such as NoV (GI & GII), Rotavirus, Sapovirus, Human Astrovirus and Human Adenovirus (van Alphen et al. 2014).

Nucleic acid sequence-based amplification (NASBA) is a sensitive and efficient isothermal transcription-based amplification method specifically designed for the detection of RNA targets. This method targets RNA at a single fixed temperature (41°C) and provides an alternative approach to the amplification of DNA sequences at varying temperatures compared to PCR (Wyn-Jones et al. 2014). In a comparison study, the detection of enteroviruses in surface water samples was slightly less sensitive in detecting target virus sequences than RT-PCR (Rutjes et al. 2005). This technique was reviewed as a method for the detection of microbial pathogens in food and environmental samples (Cook 2003). Jean and his group developed a multiplex format for hepatitis A virus, noroviruses genogroup I and genogroup II (Jean et al. 2004), which previously used the NASBA coupled to an ELISA reaction for the detection of rotavirus in seeded sewage effluent samples (Jean et al. 2002). Another study coupled NASBA to a molecular beacon for real-time detection of HAV in seeded surface water samples (Abd el-Galil et al. 2005).

3.2.1 *Oligonucleotide DNA microarrays and pyrosequencing*

This is a molecular tool for detecting mutations in DNA sequences and very helpful for characterizing pathogens in environmental samples. Microarrays also have the ability for detecting the host origin of contaminants, and for characterization of contamination in different sources of water using specific probes. Pyrosequencing could also identify novel pathogens associated with water and has been used to the detection of bacteriophages in reclaimed and potable and lake waters (Ramírez-Castillo et al. 2015).

3.3 *Molecular biology combined with cell culture*

3.3.1 *Integrated cell culture-polymerase chain reaction (ICC-PCR)*

Combination of cell culture with PCR was the practical solution to detect the infectious virus which may not produce CPE or exhibit limited growth. This technique is generally used to detect both DNA viruses (ICC-PCR) and RNA viruses (ICC/RT-PCR). Various studies showed that virus replication can be detected before it presents CPE which is considered one of the major advantages of this technique. The environmental water sample was inoculated to BGM cell culture to detect enteroviruses using this technique (Reynolds et al. 1996, Murrin and Slade 1997).

After withdrawing supernatants at intervals of up to 10 days, the virus was detectable as early as 1-day post-inoculation, instead of more than 3 days until present CPE. Different studies also showed that using ICC-PCR may increase the isolation of some enteric viruses as well as the positive sample record (Chapron et al. 2000, Lee and Jeong 2004, Dong et al. 2010). The same method was able to detect the infectious particles of enteroviruses and adenoviruses in 3 days and 5 days respectively, compared with 5 days and 10 days if plaque assays were used (Greening et al. 2002). Human adenovirus infectivity by ICC-PCR has been tested by inoculation into A549 cell cultures and subsequently the supernatants are assessed by PCR (Wyn-Jones et al. 2014). Recently, a study developed a successful method using the integrated cell culture reverse transcriptase quantitative PCR (ICC/RT-qPCR) method to specifically quantify the infectious concentrations of eight enteroviruses serotypes commonly encountered in sewage (Larivé et al. 2021).

3.3.2 *Viral messenger RNA targeting assay*

Some DNA viruses do not replicate well in cell cultures such as adenoviruses (Ad), particularly Ad40 and 41, as they do not produce a clear CPE. Therefore, assays targeting viral mRNA for the detection of human adenoviruses in environmental samples have been developed but have not been widely applied in ambient waters (Rodriguez et al. 2013). In 2003, a study developed a method to detect the infectious Ad2 and Ad41 in A549 cell culture by detecting virus-specific mRNA that is created only at the replication stage (Ko et al. 2003). In this study, mRNA of Ad2 was detected as soon as 6 hours after infection in comparison to Ad41 that was detected as soon as 24 hours after infection. In addition, the same group developed the technique and was able to detect adenoviruses in the water sample as 2 IU for Ad2 and 10 IU for Ad41 in sample concentrates inoculated into cell cultures (Cromeans et al. 2004). Otherwise, the enteric Ad40 and Ad41 viruses cannot be grown in most cell culture systems (Wyn-Jones et al. 2014).

4. Concluding remarks

Since the confirmation of viral transmission through water, virus assessment methods have also been developed to increase the efficiency of detection. Sample concentration is considered the first step in this process and therefore different techniques have been tested for various sample volumes. Aqueous polymer two-phase separation, hydroextraction, and soluble alginate ultrafilter membrane techniques have the ability to concentrate viruses from water with good recovery efficiency. But, all of them are favorable to treat just small volumes of water including the turbid water samples that need a prefiltration step. The ultracentrifugation technique can make fractions of diverse virus types which is useful in some protocols but it has an exorbitant cost disadvantage with excessive processing time. The electronegative filter is considered the commonly used technique in routine work for both small and large volumes, but it remains most ideal for small volumes of water. Despite its low-cost price as an advantage, it still requires a preconditioning procedure for the sample. The electropositive filters were developed to overcome the preconditioning step for the sample which widely supported their use in large volumes of water. However, it continues to be quite expensive, with a high possibility of clogging problems during filtration. The ultrafiltration technique has been increasingly used basically during the concentration of multiple microbial processes. While it does remain a cost efficient procedure without any preconditioning step, clogging problems and slow filtration rates are still the most noticeable disadvantages of this technique.

Assessment of the efficiency of virus recovery in water samples depends mainly on the concentration method and sequentially on the detection method used. Enteric viruses could be detected by cell culture techniques and/or molecular biology techniques. Although most enteric viruses are not adapted to cultivate in cell culture, it is considered as one of the methods used to detect viruses and measure their ability to multiply. Plaque assay and 50% tissue culture infective dose methods are commonly used to quantify virus particles in the water sample. Due to the multiplicity variation between viruses, the incubation period, as well as the method time, may extend to 12 days until the cytopathic effect is visible and the final result is obtained. Since most viruses can't produce a visible CPE, the alternative end-point dilution assays can be used where it can measure virus titer before CPE appearance. The molecular biology technique is considered the most promising technique in the waterborne virus detection process. The qualitative PCR and the quantitative real-time PCR assays are widely used to detect RNA and DNA viruses even in small quantities. The technique has the ability to target a specific virus type using a single set of primers in one condition or various virus types using a pool of primers in one condition. In addition, the NASBA method is used to produce multiple copies of single-strand RNA with a two-step process with the advantage to work at a constant temperature within a short time.

The current pandemic has stressed on the utility of using environmental and water surveillance to monitor the increase of particular pathogens in the community. Nowadays, there are wastewater surveillance systems for detecting enteric viruses or SARS-CoV-2 to prevent outbreaks. So, it is necessary to encourage environmental monitoring for preventing the spread of enteric viruses and to investigate the importance of wastewater effluents in their transmission.

Reference

Abd el-Galil, K.H., el-Sokkary, M.A., Kheira, S.M., Salazar, A.M., Yates, M.V., Chen, W. et al. 2005. Real-time nucleic acid sequence-based amplification assay for detection of hepatitis A virus. Appl. Environ. Microbiol. 71: 7113–7116.

Anderson, N.G., Cline, G.B., Harris, N.W. and Green, J.G. 1967. Isolation of viral particles from large fluid volumes. pp. 75–88. *In*: Berg, G. (ed.). Transmission of Viruses by the Water Route. London: Interscience.

Badr, K.R., Parente-Rocha, J.A., Baeza, L.C., Ficcadori, F.S., Souza, M., Soares, C.M., Guissoni, A.C.P. et al. 2019. Quantitative proteomic analysis of A549 cells infected with human adenovirus type 2. J. Med. Virol. 2019 Jul; 91(7): 1239–1249.

Baker, M.N.A.T. and Michael, J. 1981. The Quest for Pure Water: The History of the Twentieth Century: American Water Works Association - AWWA, 2. Ed.

Beer, K.D., Gargano, J.W., Roberts, V.A., Hill, V.R., Garrison, L.E., Kutty, P.K. et al. 2015a. Surveillance for Waterborne Disease Outbreaks Associated with Drinking Water—United States, 2011–2012. Morbidity and Mortality Weekly Report, August 14, 64(31): 842–848.

Beer, K.D., Gargano, J.W., Roberts, V.A., Reses, H.E. et al. 2015b. Outbreaks Associated With Environmental and Undetermined Water Exposures—United States, 2011–2012. Morbidity and Mortality Weekly Report, August 14, 2015b, 64(31): 849–851.

Bennett, H.B., O'Dell, H.D., Norton, G., Shin, G., Hsu, F.-C. and Meschke, J.S. 2010. Evaluation of a novel electropositive filter for the concentration of viruses from diverse water matrices. Water Science and Technology 61: 317–322.

Berg, G., Dahling, D.R. and Berman, D. 1971. Recovery of small quantities of viruses from clean waters in cellulose nitrate membrane filters. Applied Microbiology 22: 608–614.

Berman, D., Rohr, M.E. and Safferman, R.S. 1980. Concentration of poliovirus in water by molecular filtration. Applied and Environmental Microbiology 40: 426–428.

Blanco, A., Abid, I., Al-Otaibi, N., Pérez-Rodríguez, F.J., Fuentes, C., Guix, S. et al. 2019. Glass wool concentration optimization for the detection of enveloped and non-enveloped waterborne viruses. Food Environ. Virol. 11: 184–192.

Cashdollar, J.L. and Dahling, D.R. 2006. Evaluation of a method to re-use electropositive cartridge filters for concentrating viruses from tap and river water. J. Virol. Methods 132: 13–17.

Cashdollar, J.L. and Wymer, L. 2013. Methods for primary concentration of viruses from water samples: A review and meta-analysis of recent studies. J. Appl. Microbiol. 115: 1–11.

Chang, S.L., Stevenson, R.E., Bryant, A.R., Woodward, R.L. and Kabler, P.W. 1958. Removal of Coxsackie and bacterial viruses in water by flocculation. II. Removal of Coxsackie and bacterial viruses and the native bacteria in raw Ohio River water by flocculation with aluminum sulfate and ferric chloride. Am. J. Public Health Nations Health 48: 159–169.

Chang, S.L. 1968. Waterborne viral infections and their prevention. Bull World Health Organ. 38(3): 401–414.

Chang, L.T., Farrah, S.R. and Bitton, G. 1981. Positively charged filters for virus recovery from wastewater treatment plant effluents. Appl. Environ. Microbiol. 42: 921–924.

Chapron, C.D., Ballester, N.A., Fontaine, J.H., Frades, C.N. and Margolin, A.B. 2000. Detection of astroviruses, enteroviruses, and adenovirus types 40 and 41 in surface waters collected and evaluated by the information collection rule and an integrated cell culture-nested PCR procedure. Appl. Environ. Microbiol. 66: 2520–2525.

Cliver, D.O. 1965. Factors in the membrane filtration of enteroviruses. Applied Microbiology 13: 417–425.

Cliver, D.O. and Yeatman, J. 1965. Ultracentrifugation in the concentration and detection of enteroviruses. Applied Microbiology 13: 387–392.

Coin, L. 1967. The viruses in water. p. 367. *In*: Berg, G. (ed.). Transmission of Viruses by the Water Route. London: Interscience

Condit, R.C. 2013. Principles of virology. pp. 19–51. *In*: Fields Virology; Wolters Klewers/Lippincott Williams & Wilkins: Philadelphia, PA, USA.

Cook, N. 2003. The use of NASBA for the detection of microbial pathogens in food and environmental samples. J. Microbiol. Methods 53: 165–174.

Cromeans, T.L., Jothikumar, N., Jung, K., Ko, G., Wait, D. and Sobsey, M.D. 2004. Development of Molecular Methods to Detect Infectious Viruses in Water. Denver, CO: Awwa Research Foundation.

Dahling, D. 2002. An improved filter elution and cell culture assay procedure for evaluating public groundwater systems for culturable enteroviruses. Water Environment Research 74: 564–568.

Dahling, D.R., Berg, G. and Berman, D. 1974. BGM, a continuous cell line more sensitive than primary rhesus and African green kidney cells for the recovery of viruses from water. Health Lab Sci. 11: 275–282.

Dhopeshwarkar, G.A., Rao, K.R. and Viswanathan, R. 1957. Infectious hepatitis: biochemical studies. Indian J. Med. Res. 45: 125–133.

Diniz-Mendes, L., Paula, V.S., Liz, S.L. and Niel, C. 2008. High prevalence of human Torque Teno Virus in streams crossing the city of Manaus, Brazilian Amazon. J. Appli. Microbiol. 105(1): 51–58.

Diniz-Mendes, L. 2018. Virologia Ambiental. *In*: Virologia Humana e Veterinária. Ed. Thieme Revinter. 1. Ed. pp. 137–145.

Dong, Y., Kim, J. and Lewis, G.D. 2010. Evaluation of methodology for detection of human adenoviruses in wastewater, drinking water, stream water and recreational waters. J. Appl. Microbiol. 108: 800–809.

Dulbecco, R. 1952. Production of plaques in monolayer tissue cultures by single particles of an animal virus. Proc. Natl. Acad. Sci. USA 38(8): 747–752.

Elford, W.J. 1931a. A new series of graded collodion membranes suitable for general bacteriological use, especially in virus studies. J. Pathol. Bacteriol. 34: 505–521.

Fagnant, C.S., Toles, M., Zhou, N.A. et al. 2017. Development of an elution device for ViroCap virus filters. Environ. Monit. Assess. 189(11): 574. Published 2017 Oct 19.

Farrah, S.R. and Bitton, G. 1979. Low molecular weight substitutes for beef extract as eluents for poliovirus adsorbed to membrane filters. Can. J. Microbiol. 25: 1045–1051.

Gajardo, R., Bouchrit, N., Pinto´, R.M. and Bosch, A. 1995. Genotyping of rotaviruses isolated from sewage. Appl. Environ. Microbiol. 61: 3460–3462.

Gall, A.M., Marinas, B.J., Lu, Y. and Shisler, J.L. 2015. Waterborne viruses: a barrier to safe drinking water. PLoS Pathog. 11: e1004867.

Gartner, H. 1967. Retention and recovery of polioviruses on a soluble ultrafilter. pp. 121–127. *In*: Berg, G. (ed.). Transmission of Viruses by the Water Route. London: Interscience

Gerba, C.P., Betancourt, W.Q. and Kitajima, M. 2017. How much reduction of virus is needed for recycled water: A continuous changing need for assessment? Water Res. 108: 25–31.

Gibbons, C.D., Rodriguez, R.A., Tallon, L. and Sobsey, M.D. 2010. Evaluation of positively charged alumina nanofiber cartridge filters for the primary concentration of noroviruses, adenoviruses and male-specific coliphages from seawater. J. Appl. Microbiol. 109: 635–641.

Gibson, K.E. and Schwab, K.J. 2011. Tangential-flow ultrafiltration with integrated inhibition detection for recovery of surrogates and human pathogens from large-volume source water and finished drinking water. Appl. Environ. Microbiol. 77: 385–391.

Grabow, W.O.K. 1968. The virology of waste water treatment. Water Research 2: 675–701.

Graff, J., Ticehurst, J. and Flehmig, B. 1993. Detection of hepatitis A virus in sewage by antigen capture polymerase chain reaction. Appl. Environ. Microbiol. 59: 3165–3170.

Gravelle, C.R. and Chin, T.D.Y. 1961. Enterovirus isolations from sewage: A comparison of three methods. Journal of Infectious Disease 10: 205.

Greenberg, A. and Taras, A.M.J. (eds.). 1981. Standard Methods for the Examination of Water and Wastewater. Washington, DC: American Public Health Association.

Greening, G.E., Hewitt, J. and Lewis, G.D. 2002. Evaluation of integrated cell culture-PCR (C-PCR) for virological analysis of environmental samples. J. Appl. Microbiol. 93: 745–750.

Grinrod, J. and Cliver, D.O. 1970. A polymer two phase system adapted for virus detection. Archive fu¨r Gesamte Virusforsch 31: 365–372.

Grinstein, S., Melnick, J.L. and Wallis, C. 1970. Virus isolations from sewage and from a stream receiving effluents of sewage treatment plants. Bulletin of the World Health Organization 42: 291–296.

Haramoto, E., Katayama, H., Utagawa, E. and Ohgaki, S. 2009. Recovery of human norovirus from water by virus concentration methods. Journal of Virological Methods 160: 206–209.

Herzog, P., Drosten, C. and Müller, M.A. 2008. Plaque assay for human coronavirus NL63 using human colon carcinoma cells. Virol. J. 5: 138.

Hill, W.F., Akin, E.W. and Benton, W.H. 1971. Detection of viruses in water: A review of methods and applications. Water Research 5: 967–995.

Hill, V.R., Polaczyk, A.L., Hahn, D., Narayanan, J., Cromeans, T.L., Roberts, J.M. et al. 2005. Development of a rapid method for simultaneous recovery of diverse microbes in drinking water by ultrafiltration with sodium polyphosphate and surfactants. Applied and Environmental Microbiology 71: 6878–6884.

Hill, V.R., Polaczyk, A.L., Kahler, A.M., Cromeans, T.L., Hahn, D. and Amburgey, J.E. 2009. Comparison of hollow-fiber ultrafiltration to the USEPA VIRADEL Technique and USEPA Method 1623. Journal of Environmental Quality 38: 822–825.

Hino, S. and Miyata, H. 2007. Torque Teno Virus (TTV): current status. Rev. Med. Virol. 17(1): 45–57.

Holowecky, P.M., James, R.R., Lorch, D.P., Straka, S.E. and Lindquist, H.D.A. 2009. Evaluation of ultrafilter cartridges for a water sampling apparatus. Journal of Applied Microbiology 106: 738–747.

Ikner, L.A., Soto-Beltran, M. and Bright, K.R. 2011. New method using a positively charged microporous filter and ultrafiltration for concentration of viruses from tap water. Appl. Environ. Microbiol. 77: 3500–3506.

Ikner, L.A., Gerba, C.P. and Bright, K.R. 2012. Concentration and recovery of viruses from water: A comprehensive review. Food Environ. Virol. 4: 41–67.

Ishii, S., Kitamura, G., Segawa, T., Kobayashi, A., Miura, T., Sano, D. and Okabe, S. 2014. Microfluidic quantitative PCR for simultaneous quantification of multiple viruses in environmental water samples. Appl. Environ. Microbiol. 80: 7505–7511.

Jean, J., Blais, B., Darveau, A. and Fliss, I. 2002. Rapid detection of human rotavirus using colorimetric nucleic acid sequence-based amplification (NASBA)-enzyme-linked immunosorbent assay in sewage treatment effluent. FEMS Microbiol. Lett. 23: 143–147.

Jean, J., D'Souza, D.H. and Jaykus, L.A. 2004. Multiplex nucleic acid sequence-based amplification for simultaneous detection of several enteric viruses in model ready-to-eat foods. Appl. Environ. Microbiol. 70(11): 6603–6610.

Johnson, J.H., Fields, J.E. and Darlington, W.A. 1967. Removing viruses from water by polyelectrolytes. Nature 213: 665–667.

Jothikumar, N., Cromeans, T.L., Hill, V.R., Lu, X., Sobsey, M.D. and Erdman, D.D. 2005. Quantitative real-time PCR assays for detection of human adenoviruses and identification of serotypes 40 and 41. Appl. Environ. Microbiol. 71(6): 3131–3136.

Jothikumar, N., Kang, G. and Hill, V.R. 2009. Broadly reactive TaqMan assay for real-time RT-PCR detection of rotavirus in clinical and environmental samples. JIN2@cdc.gov. J. Virol. Methods Feb; 155(2): 126–31.

Kulinkina, A.V., Shinee, E., Herrador, B.R.G., Nygård, K. and Schmoll, O. 2016. The Situation of Water-Related Infectious Diseases in the Pan-European Region. World Health Organization, 42p.

Karim, M.R., Rhodes, E.R., Brinkman, N., Wymer, L. and Fout, G.S. 2009. New electropositive filter for concentrating enteroviruses and noroviruses from large volumes of water. Applied and Environmental Microbiology 75: 2393–2399.

Kelly, S.M. 1953. Detection and occurrence of Coxsackie viruses in sewage. Am. J. Public Health Nations Health 43: 1532–1538.

Kern, A., Kadar, M., Szomor, K., Berencsi, G., Kapusinszky, B. and Vargha, M. 2013. Detection of enteric viruses in Hungarian surface waters: first steps towards environmental surveillance. J. Water Health 11(4): 772–82. doi:10.2166/wh.2013.242.

Ko, G., Cromeans, T.L. and Sobsey, M.D. 2003. Detection of infectious adenovirus in cell culture by mRNA reverse transcription-PCR. Appl. Environ. Microbiol. 69: 7377–7384.

Koonin, E.V. and Starokadomskyy, P. 2016. Are viruses alive? The replicator paradigm sheds decisive light on an old but misguided question. Studies in History and Philosophy of Biological and Biomedical Sciences 59: 125–34.

La Rosa, G., Pourshaban, M., Iaconelli, M. and Muscillo, M. 2010. Quantitative real-time PCR of enteric viruses in influent and effluent samples from wastewater treatment plants in Italy. Ann. Ist Super Sanita 46(3): 266–73.

Lambertini, E., Spencer, S.K., Bertz, P.D., Loge, F.J., Kieke, B.A. and Borchardt, M.D. 2008. Concentration of enteroviruses, adenoviruses, and noroviruses from drinking water by use of glass wool filters. Applied and Environmental Microbiology 74: 2990–2996.

Landsteiner, K. and Popper, E. 1909. Übertragung der Poliomyelitis acuta auf Affen. Z Immunitätsforsch 2: 377–390.

Larivé, O., Brandani, J., Dubey, M. and Kohn, T. 2021. An integrated cell culture reverse transcriptase quantitative PCR (ICC-RTqPCR) method to simultaneously quantify the infectious concentrations of eight environmentally relevant enterovirus serotypes. J. Virol. Methods 296: 114225.

Laverick, M.A., Wyn-Jones, A.P. and Carter, M.J. 2004. Quantitative RT-PCR for the enumeration of noroviruses (Norwalk-like viruses) in water and sewage. Lett. Appl. Microbiol. 39: 127–136.

Le Cann, P., Ranarijaona, S., Monpoeho, S., Le Guyader, F. and Ferre, V. 2004. Quantification of human astroviruses in sewage using real-time RT-PCR. Res. Microbiol. 155: 11–15.

Lee, H.K. and Jeong, Y.S. 2004. Comparison of total culturable virus assay and multiplex integrated cell culture-PCR for reliability of waterborne virus detection. Appl. Environ. Microbiol. 70: 3632–3636.

Leifels, M., Hamza, I.A., Krieger, M., Wilhelm, M., Mackowiak, M. and Jurzik, L. 2016. From lab to lake—evaluation of current molecular methods for the detection of infectious enteric viruses in complex water matrices in an urban area. PLoS ONE 11: e0167105.

Leifels, M., Shoults, D., Wiedemeyer, A., Ashbolt, N., Sozzi, E., Hagemeier, A. and Jurzik, L. 2019. Capsid integrity qPCR-An Azo-Dye based and culture-independent approach to estimate adenovirus infectivity after disinfection and in the aquatic environment. Water 11: 1196.

Leland, D.S. and Ginocchio, C.C. 2007. Role of cell culture for virus detection in the age of technology. Clin. Microbiol. Rev. 20: 49–78.

Lodder, W.J. and de Roda Husman, A.M. 2005. Presence of noroviruses and other enteric viruses in sewage and surface waters in The Netherlands. Appl. Environ. Microbiol. 71: 1453–1461.

Marx, F.E., Taylor, M.B. and Grabow, W.O.K. 1995. Optimization of a PCR method for the detection of astrovirus type 1 in environmental samples. Water Sci. Technol. 31: 359–362.

Melnick, J.L., Emmons, J., Coffey, J.H. and Schoof, H. 1954. Coxsackie viruses from sewage. American Journal of Hygiene 59: 185–195.

Metcalf, T.G. 1961. Use of membrane filters to facilitate the recovery of virus from aqueous suspensions. Applied Microbiology 9: 376–379.

Michen, B. and Graule, T. 2010. Isoelectric points of viruses. J. Appl. Microbiol. 109: 388–397.

Morris, R. 1985. Detection of enteroviruses: an assessment of ten cell lines. Water Sci. Technol. 17: 81–88.

Murrin, K. and Slade, J. 1997. Rapid detection of viable enteroviruses in water by tissue culture and semi-nested polymerase chain reaction. Water Sci. Technol. 35: 429–432.

Naidu, S.S. and Viswanathan, R. 1957. Infectious hepatitis in pregnancy during Delhi epidemic. Indian J. Med. Res. 45: 71–76.

Nupen, E.M. 1970. Virus studies on the Windhoek waste-water reclamation plant (South Africa). Water Research 4: 661–672.

Olszewski, J., Winona, L. and Oshima, K.H. 2005. Comparison of 2 ultrafiltration systems for the concentration of seeded viruses from environmental waters. Canadian Journal of Microbiology 51: 295–303.

Okoh, A.I., Sibanda, T. and Gusha, S.S. 2010. Inadequately treated wastewater as a source of human enteric viruses in the environment. Int. J. Environ. Res. Public Health 7: 2620–2637; doi:10.3390/ijerph7062620.

Pádraig Strappe. 1991. Rotavirus detection—a problem that needs concentration. Water Sci. Technol. 1 July; 24(2): 221–223.

Patel, J.R., Daniel, J. and Mathan, V.I. 1985. A comparison of the susceptibility of three human gut tumour-derived differentiated epithelial cell lines, primary monkey kidney cells and human rhabdomyosarcoma cell line to 66-prototype strains of human enteroviruses. J. Virol. Methods 12: 209–216.

Polaczyk, A.L., Roberts, J.M. and Hill, V.R. 2007. Evaluation of 1MDS electropositive microfilters for simultaneous recovery of multiple microbe classes from tap water. J. Microbiol. Methods Feb; 68(2): 260–6.

Puig, M., Jofre, J., Lucena, F., Allard, A., Wadell, G. and Girones, R. 1994. Detection of adenoviruses and enteroviruses in polluted water by nested PCR amplification. Appl. Environ. Microbiol. 60: 2963–2970.

Ramírez-Castillo. 2015. Waterborne pathogens: detection methods and challenges. Pathogens 4: 307–334; doi:10.3390/pathogens4020307.

Randazzo, W., Vasquez-Garcia, A., Aznar, R. and Sanchez, G. 2018. Viability RT-qPCR to distinguish between HEV and HAV with intact and altered capsids. Front. Microbiol. 9: 1973.

Rao, N.U. and Labzoffsky, N.A. 1969. A simple method for the detection of low concentration of viruses in large volumes of water by the membrane filtration technique. Canadian Journal of Microbiology 15: 399–403.

Reed, L.J. and Muench, H. 1938. A simple method of estimating fifty percent endpoints. Am. J. Hyg. 27: 493–495.

Reynolds, C.A., Gerba, C.P. and Pepper, I.L. 1996. Detection of infectious enterovirus by an integrated cell culture-PCR procedure. Appl. Environ. Microbiol. 62: 1424–1427.

Rodriguez, R.A., Polston, P.M., Wu, M.J., Wu, J. and Sobsey, M.D. 2013. An improved infectivity assay combining cell culture with real-time PCR for rapid quantification of human adenoviruses 41 and semi-quantification of human adenovirus in sewage. Water Res. 47: 3183–3191.

Rutjes, S.A., Italiaander, R., van den Berg, H.H., Lodder, W.J. and de Roda Husman, A.M. 2005. Isolation and detection of enterovirus RNA from large-volume water samples by using the NucliSens miniMAG system and real-time nucleic acid sequence-based amplification. Appl. Environ. Microbiol. 71: 3734–3740.

Rybicki, E.P. 1990. The classification of organisms at the edge of life, or problems with virus systematics. South African Journal of Science 86: 182–86.

Schultz, A.C., Perelle, S., Di Pasquale, S., Kovac, K., De Medici, D., Fach, P. et al. 2011. Collaborative validation of a rapid method for efficient virus concentration in bottled water. International Journal of Food Microbiology 145: S158–S166.

Sedmak, G., Bina, D., MacDonald, J. and Couillard, L. 2005. Nine-year study of the occurrence of culturable viruses in source water for two drinking water treatment plants and the influent and effluent of a wastewater treatment plant in Milwaukee, Wisconsin (August 1994 through July 2003). Appl. Environ. Microbiol. 71: 1042–1050.

Shuval, H.J., Cymbalista, S., Fatal, B. and Goldblum, N. 1967. Concentration of enteric viruses in water by hydro-extraction and two phase separation. pp. 45–55. *In*: Berg, G. (ed.). Transmission of Viruses by the Water Route. London: Interscience.

Shuval, H.J., Fattal, B., Cymbalista, S. and Goldblum, N. 1969. The phase-separation method for the concentration and detection of viruses in water. Water Research 3: 225–240.

Sobsey, M.D. and Glass, J.S. 1980. Poliovirus concentration from tap water with electropositive adsorbent filters. Applied and Environmental Microbiology 40: 201–210.

Standing Committee of Analysts (SCA). 1995. Methods for the isolation and identification of human enteric viruses from waters and associated materials. Methods for the Examination of Waters and Associated Materials. London: HMSO.

Stevenson, R. 1967. Quantitative recovery of viruses from dilute suspensions. p. 143. *In*: Berg, G. (ed.). Transmission of Viruses by the Water Route. London: Interscience.

Toomey, J.A. 1945. Poliomyelitis. Arch. Surg. 51: 306–309.

United States Environmental Protection Agency (USEPA). 2007. USEPA/APHA Standard methods for the examination of water and wastewater.

Van Alphen, L.B., Dorléans, F., Schultz, A.C., Fonager, J., Ethelberg, S., Dalgaard, C. et al. 2014. The application of new molecular methods in the investigation of a waterborne outbreak of norovirus in Denmark, 2012. PloS One 9(9): e105053. doi:10.1371/journal.pone.0105053.

Ver, B.A., Melnick, J.L. and Wallis, C. 1968. Efficient filtration and sizing of viruses with membrane filters. J. Virol. 2: 21–25.

Victoria, M., Guimaraes, F., Fumian, T., Ferreira, F., Vieira, C., Leite, J.P. et al. 2009 Evaluation of an adsorption-elution method for detection of astrovirus and norovirus in environmental waters. J. Virol. Methods 156: 73–76.

Vilagines, P., Sarrette, B., Champsaur, H., Hugues, B., Dubrou, S., Joret, J.C. et al. 1997. Round robin investigation of glass wool method for poliovirus recovery from drinking water and sea water. Water Sci. Technol. 35: 445–449.

Viswanathan, R. and Sidhu, A.S. 1957. Infectious hepatitis; clinical findings. Indian J. Med. Res. 45: 49–58.

Wallis, C. and Melnick, J.L. 1967a. Concentration of enteroviruses on membrane filters. J. Virol. 1: 472–477.

WHO. 1979. Human viruses in water, wastewater, and soil. *In*: GROUP, W.H.O.S. (ed.). Geneva: World Health Organization.

Wyn-Jones, A.P. and Sellwood, J. 2001. Enteric viruses in the aquatic environment. J. Appl. Microbiol. 91: 945–962.

Wyn-Jones, A.P., Carducci, A., Cook, N., D'Agostino, M., Divizia, M., Fleischer, J. et al. 2011. Surveillance of adenoviruses and noroviruses in European recreational waters. Water Res. 45(3): 1025–38. doi:10.1016/j. watres.2010.10.015.

Chapter 7

Management of Microbiological Contaminants in Shellfish Growing Areas
Current Practices and Future Directions with a Focus on Viruses

Carlos J.A. Campos

1. Introduction

1.1 Evidence for the problem

Bivalve shellfish (clams, cockles, oysters, mussels) feed by filtering large volumes of water in their natural environment. They "pump" hundreds of litres of water per day into their mantle cavity and extract the particles that they use as food (Galtsoff 1964). During filter feeding, shellfish can accumulate pathogens and other contaminants from their growing waters. Contaminated shellfish can cause illness of varying severity and duration, particularly if consumed raw or lightly cooked (Rippey 1994, Butt et al. 2004). Shellfish-related illness has been recognised since the late 19th century (Dodgson 1928, Lees 2000, Reid and Durance 2001). In the past, the most common illnesses were typhoid and gastroenteritis of bacterial origin (Rippey 1994). However, the epidemiological profile has changed as more countries implement sanitary controls targeting both growing areas and processing establishments. Currently, there is a large amount of epidemiological evidence indicating that, in the developed world, illness cases of known aetiology are predominantly caused by human noroviruses (NoV) and Hepatitis A (Bellou et al. 2013). The literature contains abundant case and outbreak reports of NoV infection associated with the

Water Quality Scientist, Healthy Oceans Group, Cawthron Institute, New Zealand.
Email: carlos.campos@cawthron.org.nz

consumption of shellfish (reviewed by Richards 1985, EFSA 2012, Bellou et al. 2013). Such outbreaks occur on a regular basis worldwide (Richards 2016, Meghnath et al. 2019, Wan et al. 2018). Outbreaks of NoV occur throughout the year although there is increased activity during the winter months, at least in temperate climates where most epidemiological data are available (Ahmed et al. 2013). Illnesses due to bacteria such as vibrios, *Salmonella* spp., *Campylobacter* spp. and *Shigella* spp. do occur in some geographical areas (Iwamoto et al. 2010). Symptoms of NoV gastroenteritis usually start within 1–2 days of eating the contaminated shellfish, but may begin in as few as 12 hours (Hall et al. 2011). Projectile vomiting is often the first symptom, along with diarrhea and cramps (Cardemil et al. 2017). Headache, mild fever, and muscle aches may also occur. Most patients experience a brief, self-limited infection with symptoms resolving within 2–3 days (Cardemil et al. 2019). A number of factors affect susceptibility of an individual to these viruses, including age, NoV genotype, the route of transmission, and immuno-competence and immunity of infected individuals (Cardemil et al. 2019).

In many parts of the world, shellfish growing areas are located in shallow estuaries and other coastal environments near densely populated areas that produce large volumes of wastewater. This presents a significant management challenge to the shellfish industry and regulators because shellfish farming requires good water quality and targeted measures to control faecal contamination. Episodes of poor water quality caused by discharges of untreated or partially treated wastewater are an important barrier to further expansion of the shellfish farming sector (Brown et al. 2020, Webber et al. 2021). This chapter reviews current practices on the management of microbiological risks in shellfish growing areas. Science in this area has evolved much in recent decades and there is a need to update and introduce new scientific developments into the principles and practices of microbiological risk management of shellfish harvested for human consumption.

Risk management is an element of the risk analysis framework that considers policies, prevention and control measures and other factors to promote consumer health protection and fair trade (FAO and WHO 2021). This includes the management of risks in the face of uncertainty, balancing consideration of the actual/potential microbiological hazards with available treatment and mitigation strategies as well as the resources needed to implement the controls. While the framework for assessing microbiological risks in shellfish considers the entire farm-to-fork system, it is focused on preventing contamination in growing areas rather than on end-product testing which only highlights potential health problems after the shellfish have been consumed (FAO and WHO 2020). Following this introductory section, this chapter summarises the main elements of the current regulatory framework that include, sanitary surveys and monitoring and classification of growing areas.

1.2 Estimating health risks in growing areas

NoV is often detected in untreated and treated wastewater throughout the year, the peaks of prevalence occurring during the autumn-winter period (Katayama et al. 2008, Nordgren et al. 2009, Fumian et al. 2019). Similar seasonality is detected in shellfish growing areas. In the UK, average levels of NoV in Pacific oysters (*C. gigas*) during colder months (October–March) were found to be 17 times higher than those during the remainder of the year (Lowther et al. 2008). In England and Wales, oysters from cold waters (< 5°C) had significantly higher NoV concentrations than those from warmer waters (> 10°C) and this was assumed to be associated with the seasonality of the virus in the community and/or the metabolism of the animals (Campos et al. 2017). These authors studied the relationships between NoV concentrations in oysters from 31 commercial growing areas and demographic, hydrometric, climatic and pollution source characteristics of upstream river catchments. The results indicated that seawater temperature, river flows, catchment area, volume of wastewater discharges, and number of sewer overflows are key risk factors for NoV contamination in the oysters. Linear regression modelling indicated that an average of 10 wastewater spills/year or less would correspond to NoV concentrations < 100 copies/g (GI+GII) in the growing areas.

   Different profiles of NoV illness are associated with different shellfish species. The highest number of cases is associated with oysters because these species are usually consumed whole and raw. Illness cases linked to species that are traditionally cooked before consumption, such as mussels and clams, are less frequent (Lee and Younger 2002). Species that are eviscerated (e.g., scallops) prior to sale present lower risk of infection because NoV ingested by shellfish is predominantly accumulated in the digestive tissues (Wang et al. 2008). In the UK, consumption of contaminated food is estimated to account for 3–11% of NoV cases (ca. 74,000 cases/year), of which 16% are attributed to consumption of oysters (11,800 cases/year) (Hassard et al. 2017). In 2009, an outbreak of NoV gastroenteritis affected at least 240 persons who had eaten at a gourmet restaurant over a period of seven weeks in England. The ongoing risk was attributed to persistent contamination of the oyster supply alone or in combination with further spread via infected food handlers or the restaurant environment (Smith et al. 2012). The burden of viral illness associated with shellfish consumption is not only associated with morbidity and mortality factors, but can also adversely affect consumers' confidence in shellfish products. In Australia, the economic burden of NoV illness associated with contaminated oysters was estimated at $11.85 million/year (Hudson 2011). Polymerase chain reaction (PCR) methods detect viral genome and therefore does not discriminate between positive results caused by infectious viruses and those caused by non-infectious remnants (damaged virus particles, naked RNA). A number of alternative/complementary approaches to PCR testing have been proposed, including the use of infectious viral indicator organisms such as F-specific RNA bacteriophage. Lowther et al. (2019) proposed combining RT-qPCR testing with the infectivity test for F-RNA phage to better estimate actual health risks and better predict the presence of infectious NoV to avoid unproductive restrictions on producers.

Quantitative microbial risk assessment (QMRA) has been used to estimate health risks associated with various pathogens in shellfish growing areas, providing a practical alternative to expensive epidemiological investigations (WHO 2016). QMRA has also been used to investigate the relative health risks from different sources of fecal pollution, including human sewage sources, stormwater runoff, bird contamination issues, and agricultural faecal inputs. A QMRA comprises:

- Problem formulation or hazard identification to determine the pathogens and human health outcomes of concern. The pathogens most often included are enteric viruses (a group of unrelated viruses transmitted through the faecal-oral route; includes NoV and Hepatitis A) (Pintó et al. 2009).

- Exposure assessment to determine the pathways of exposure and quantify the pathogen exposure doses for selected events. Site-specific quantification of pathogen concentrations is preferred over modelled concentrations because they result in more empirical estimates of the exposure dose.

- Dose-response assessment to determine the relationships between the exposure dose and the likelihood of the health outcome.

- Risk characterization to generate a quantitative measure of risk based on the outcomes of the exposure assessment and dose-response assessment.

1.3 Sources of microbiological contamination

Microbiological contamination of shellfish growing areas originates from a variety of sources and under many different environmental conditions. Generally, sources of contamination are divided into two categories: point and non-point. A point source enters the receiving water at discrete, measurable locations such as discharge of wastewater from treatment plants, pump stations, septic tanks and industries. Non-point sources of contamination refer to human activities and natural processes in the catchment which are diffuse or dispersed and therefore are difficult to define or quantify. These include urban runoff, agricultural runoff, wastewater discharges from boats, faecal matter from wildlife, etc. (Garreis 1994, Campos et al. 2013).

Concentrations of faecal indicator bacteria (FIB) (faecal coliforms or *Escherichia coli*) in nearshore waters often correlate significantly with human population and percentage of urban area in upstream catchments (Mallin et al. 2001, van Dolah et al. 2008, Garbossa et al. 2017). Catchments with more than 10% impervious surface tend to elevate concentrations of indicator bacteria in downstream shellfish growing areas (Mallin et al. 2001, Campos and Cachola 2007). van Dolah et al. (2008) found that 62% of monitoring sites in South Carolina estuaries with more than 50% urban cover exceeded the regulatory faecal coliform standard for shellfish growing areas. In the UK, improved grassland and associated livestock are significant sources of FIB during high river flows while during base river flows, urban (sewerage-related) sources contribute the larger proportion of bacterial concentrations when there is little or no runoff from agricultural land (Kay et al. 2010). In urban catchments, the largest WWTPs can contribute > 90% of the total bacterial load to nearshore waters during dry weather (Wither et al. 2005). During wet-weather, combined sewer overflows (CSOs) can also contribute high concentrations of FIB and enteric viruses (Hata et al. 2014, Campos et al. 2016). Elevated bacterial concentrations may also be found during the summer holiday season in areas with large variation in human population due to tourism (Guillon-Cottard et al. 1998, Campos and Cachola 2007). The spatial and temporal variabilities of microbiological contamination are also determined by rainfall events and other catchment characteristics (hydrogeology, topography, rivers network). Several studies have found significant direct correlations between concentrations of *E. coli* in water/shellfish and rainfall intensity and duration and/or river flows (Campos et al. 2011, Colaiuda et al. 2021).

1.3.1 Microbial source tracking

While elevated concentrations of FIB indicate that pathogens may be present in the environment, FIB monitoring does not provide information on the source(s) of contamination. This information is critical for informing sanitary survey assessments, developing sampling plans, and enabling effective management of shellfish growing areas. Microbial source tracking (MST) is an expanding area of research that uses non-biological and/or biological markers to discriminate (track) human and non-human sources of faecal contamination (Tetra Tech and Herrera Environmental Consultants 2011). The source of host-associated microbes are identified from a range of Polymerase Chain Reaction (PCR)-based markers that target host-specific (found in only one host species or group) or host-associated (largely confined to one host species or group) indicator organisms (such as viruses, bacteria, and bacteriophages). MST methods are commonly divided into:

- Library-dependent which identifies faecal sources from water samples based on databases or "libraries" of genotypic or phenotypic fingerprints for bacterial strains of known faecal sources.
- Library-independent which identify sources based on known host-specific characteristics of the bacteria or virus and therefore do not require a library.

Library-dependent methods tend to be more expensive and require more time as well as the use of experienced personnel to complete the analysis due to the time it takes to develop a library. Additionally, libraries tend to be specific to a defined geographical area. While this level of specificity can be useful for a specific study, they are generally not as applicable on a national scale. The main advantage of library-dependent methods is that they can identify multiple sources (e.g., human, livestock, wildlife) of indicator bacteria. Recent advances in gene chip, microarray, and biosensor technologies have allowed greater sensitivity and specificity of detection as well as detection of multiple organisms or molecular markers with a single assay (Scott et al. 2002).

There is no single MST method that can identify specific sources with absolute certainty. Therefore, MST studies should be planned to answer specific questions after thorough analysis of sanitary survey or microbiological monitoring results and should address a list of prioritised study goals and objectives (Tetra Tech and Herrera Environmental Consultants 2011). Ideally, an MST study should also include multiple sampling sites, collection of multiple samples during a sampling event, and multiple sampling events during the period(s) of interest (e.g., monthly, seasonal) or hydrological conditions (e.g., base-flow and storm-flow conditions). The level of discrimination required for an MST study may be to determine the presence or relative contribution of faecal sources of human origin, determine the relative contribution of non-human sources or determine the contribution of all faecal sources (Simpson et al. 2002). MST testing should be used to complement rather than substitute other methods and tools for evaluating and identifying contaminant sources such as sanitary surveys, monitoring of FIB and water quality investigations. MST results will likely be most useful to confirm the presence of a particular source in a specific area of the catchment or to obtain a better understanding of the contribution of individual sources to the total microbiological loading from the (sub-)catchment (Tetra Tech and Herrera Environmental Consultants 2011). Verhougstraete et al. (2015) studied 64 rivers that drain 84% of the Lower Peninsula in the Great Lakes (USA) during base flow conditions for concentrations of *Escherichia coli*, *Bacteroides thetaiotaomicron* (a human source tracking marker) and the landscape, geochemical and hydrological characteristics of the catchments. This study found that septic tank discharges were the main source of faecal contamination affecting water quality in the rivers and that catchments with > 1,621 septic tanks had significantly higher concentrations of the human marker. Because MST targets can vary geographically, source specificity and sensitivity of different targets should be evaluated when selecting the most suitable target(s) for use as part of the sanitary survey or microbiological monitoring programme (Gyawali and Hewitt 2020). For greater confidence in the results, multiple markers of different phylogenies should be used (Gyawali and Hewitt 2020). While markers exist for domestic animals (cattle, poultry, horses, pigs, dogs) the distribution and performance of animal markers are not as well understood. In particular, markers have not been developed for many domestic animals that are important contributors to faecal loading in surface waters (Harwood et al. 2014). Concerning the use of MST data in health risk assessments, Zhang et al. (2019) discuss the combination of QMRA and MST to provide additional pathogen-related information for prioritising and addressing health risks that is not commonly provided by conventional monitoring of FIB. The authors consider that QMRA estimates can be improved by incorporating decay rates, recovery efficiencies, and by increasing knowledge regarding the proportion of pathogens that are viable or infectious.

2. Fundamental control measures

2.1 Sanitary surveys

Sanitary surveys (sanitary profiling) is an assessment tool that evolved from the hazard analysis and critical control point (HACCP) framework and provides a measure of the risk of microbiological contamination in SPAs (Lee et al. 2010). A sanitary survey commonly comprises:

a. an evaluation of the sources of faecal contamination that may affect the sanitary quality of the shellfish growing areas
b. an evaluation of the meteorological and hydrographic factors that may affect abundance and distribution of microbiological contaminants in the growing area
c. an assessment of the microbiological quality of the waters/shellfish.

To inform the survey, desktop reviews of the available information on pollution sources of human and animal origins are undertaken together with site visits of the shoreline and upstream areas of the catchment to visually confirm the presence of the pollution sources identified in the desktop review (Kershaw et al. 2012). Site visits may require the collection of water/shellfish samples for microbiological testing, especially if the wastewater discharges are operating at the time of the survey. The output of a sanitary survey is a report containing the results of a–c, including the microbiological results and the contextual information on pollution sources identified in the site visit (Cefas 2018, USFDA and ISSC 2019).

The main sources of faecal contamination are wastewater discharges (e.g., municipal WWTPs, storm overflows, industrial discharges, septic tanks) and non-point sources (e.g., waste discharges from boats, stormwater runoff, areas with high densities of wild animals). The level of microbiological risk varies between sources and depends upon the type, frequency and volume of the discharges. For instance, the risk of contamination from boat discharges can be reduced through installation of wastewater treatment systems on board, provision of pump-out facilities at marinas or ports and/or designation of no discharge zones in the coastal marine environment. The risk of contamination from septic tanks is usually low but can increase significantly when the soil type is unsuitable, the drainage field is incorrectly installed or the system is overloaded.

Municipal and industrial wastewater discharges should be described in the survey in terms of proximity to the shellfish growing areas, treatment level, design capacity versus actual loading, type and concentration of contaminants discharged, and discharge frequency and volume. To determine the effect of each source on the growing areas, the fate and transport of microbiological contaminants should be studied. This information can be obtained from analysis of tidal charts or stream software, simple dilution modelling or more complex particle tracking and water quality hydrodynamic modelling (Ao and Goblick 2016, Cefas 2018, García-García et al. 2021).

Municipal WWTP discharges are the contamination inputs of primary interest for shellfish growing area management because they have the highest probability of containing viral pathogens (Pouillot et al. 2022). Field data describing typical concentrations of FIB and enteric pathogens in untreated and treated wastewater effluents and microbiological removal efficiencies in wastewater treatment processes are available in the literature (Kay et al. 2008, Pouillot et al. 2015, Campos et al. 2016). These microbiological data, combined with flow measurements, can be used to estimate the microbiological loading associated with each discharge using source apportionment modelling. Methods based on source apportionment theory have been used to estimate mean FIB flux from potential point and diffuse sources during both low and high flow conditions (Stapleton et al. 2008, 2011). These source apportionment studies demonstrate the importance of high flow conditions in driving the bacterial loading from catchment pollution sources to shellfish growing areas. Generally, during high flow conditions, runoff from farmland and associated livestock is the predominant 'source' of FIB, while during low flows, when there is little or no runoff from farmland, sewage-related sources dominate bacterial loading in both rural and urban catchments. Graphical display of microbiological loads and flow volumes from individual sources at catchment level provides good representation of where investment should be made concerning wastewater treatment and risk management strategies. Catchment-scale models to simulate fate and transport of enteric viruses are still in early stages of development because of the challenges associated with targeted sampling needed to 'capture' the more episodic flux of viruses in the environment relative to FIB.

A number of process-based modelling platforms are available to simulate the fate and transport of microbiological contaminants at catchment-scale (e.g., SWAT, HSPF, WAM, QMRAcatch). The characteristics of these models vary considerably, namely their ability to represent hydrological processes (surface runoff) which in turn need to account for variations in land use, topography, soil type, vegetation cover, precipitation and land management practices (e.g., manure applications, livestock grazing), temporal and spatial resolutions, and ability to represent in-stream sources and sinks and microbiological die-off and regrowth (Cho et al. 2016). Modelling outputs are extremely valuable to help make informed decisions on appropriate pollution reduction strategies and investment to reduce faecal contaminant fluxes. For discharge receiving environments, there are many types of models for simulating the dispersion and dilution of wastewater discharges. These models also vary considerably in type and complexity. From the outset, it is important to consider how the model selected meets the requirements of the modelled scenario(s). Other initial considerations when selecting a model to simulate the mixing and transport of wastewater discharges are the type of contaminant(s) modelled, the characteristics of the receiving environment, and the region of interest for modelling (near-field or far-field) (Oliver et al. 2016, Campos et al. 2022). Near-field mixing relates to oceanographic processes that occur at distances in the order of tens to hundreds metres from the outflow jet and at temporal scales ranging from seconds to minutes. Popular near-field mixing models are CORMIX, Visual Plumes, and Proteus. Far-field mixing relates to processes that occur at temporal scales ranging from several hours to days and at distances in the order of hundreds of metres to tens of kilometres from the discharge point. Commonly used far-field models are TELEMAC, DELFT3D, MIKE21 and EFDC. The resources required to set up these models also vary considerably and depend a lot on the available data (bathymetry, tidal harmonics or currents, water levels, salinities, temperature, information on pollution sources, etc.).

Discharges of sewage from municipal and private WWTPs (Nordgren et al. 2009, Rajko-Nenow et al. 2013), overflows from sewerage systems (Rodríguez et al. 2012) and septic tanks (Borchardt et al. 2011) can introduce high quantities of NoV into shellfish growing areas. Catchments at higher risk of contamination are those with combined sewerage infrastructure characterised by large variations in flow and contaminant concentrations with the associated requirement to discharge storm overflows without treatment. In an urban catchment in Chicago (USA), average levels of NoV in water samples collected near CSO discharges increased > 10 times during wet weather (Rodríguez et al. 2012). Secondary-treated effluents are often contaminated with NoV and viral concentrations, in general, reflect the viruses circulating in the local community regardless of the symptoms shown by the contributing population (Iwai et al. 2009). Many studies have shown that primary and secondary treatment processes are less efficient at removing enteric viruses than FIB (e.g., Payment et al. 2001, Ottoson et al. 2006, Campos et al. 2016). Concerning ultra-violet (UV) disinfection, the rate of virus inactivation depends upon the UV dose absorbed by the viruses and their resistance to UV.

Rockey et al. (2020) used a PCR-based approach to accurately track positive-sense single-stranded RNA ((+) ssRNA) virus removal and inactivation. They firstly confirmed the validity of the approach with a culturable positive-sense single-stranded RNA human virus, coxsackievirus B_5, by applying both qPCR- and culture-based methods to measure inactivation kinetics with UV_{254} treatment. The authors then applied the qPCR-based method to establish a UV_{254} inactivation curve for NoV and found a constant inactivation rate (0.27 cm^2 mJ^{-1}). The results of this study suggested that NoV behaves similarly to other enteric (+) ssRNA viruses through UV_{254} inactivation, although bacteriophage MS2 is more resistant due to its shorter genome length.

Low reductions (0.4–0.8 $\log_{10}$) have been reported for NoV (GI + GII) in full scale operational UV treatment plants when the virus was exposed to UV doses of 30–40 mJ/cm^2 (Simhon et al. 2019). Membrane bioreactor technology has been increasingly implemented in areas that require high quality effluents, including those near shellfish growing areas. This treatment process combines a suspended growth biological reactor with solids removal via a filtration process. Average NoV removal rates through MBR reported in the literature are quite variable, ranging from 1 to 4 $\log_{10}$ (Ottoson et al. 2006, Sima et al. 2011, Simmons et al. 2011).

Near-field mixing is the most relevant for small volume wastewater spills such as those occurring from septic tanks or CSOs. To ensure that model outputs are accurate and reliable, it is important that the hydrodynamic model is properly calibrated and validated with field data. For areas with little or no water quality data available, new data can be collected through dedicated water quality surveys and/or tracer studies. Chemical tracers such as fluorescent dyes (e.g., rhodamine WT, fluorescein), artificial sweeteners, fluorescent whitening agents, sterols and stanols (Lim et al. 2017) or microbiological tracers (e.g., phages of *Serratia* or *Bacillus* spores, *Enterobacter*, and MS2) (Wyer et al. 2010) have been used to track wastewater discharges either visually or quantitatively. Synthetic double stranded DNA tracers show promising results for tracking wastewater discharges in soils and groundwater but further studies are needed to investigate the interactions between DNA and effluents and the impact of subsurface environmental conditions on DNA attenuation (Pang et al. 2017). Rhodamine WT is frequently used in shellfish water quality studies (Goblick et al. 2011, 2016, Campos et al. 2017). This tracer disperses readily, is very stable and non-toxic at low concentrations, and is easily detected in the marine environment using submersible fluorometers. The use of a conservative tracer such as rhodamine WT is an appropriate option to simulate NoV transport because of the long environmental persistence of these viruses (Campos et al. 2017).

Dye tracing studies involve considerably more resources than desk-based dilution modelling but overcome many of the drawbacks of simple dilution estimations. While a dye study represents a 'real-world' situation, it can be prohibitively expensive to

Table 1. Examples of norovirus contamination impact in coastal waters reported in the literature.

Study site	Type of sample	Area of norovirus impact from the discharge point	Receiving environment	Reference
Canada (Northwest Pacific)	Pacific oysters (*C. gigas*)	< 7 km	Coastal sound	Green et al. (2022)
Australia (New South Wales)	Oysters (*S. glomerata*)	4 $\log_{10}$ reduction over ≈ 5 km. Mean concentration 260 copies/g (GII)	Shallow estuary	Brake et al. (2018)
Wales (Kinmel Bay)	Mussels (*M. edulis*)	≈ 4 $\log_{10}$ reduction over 1 km (GII)	Open coast environment	Winterbourn et al. (2016)
England	Oysters (*O. edulis*)	< 1 $\log_{10}$ reduction over 12 km. Mean concentration of 1,000 copies/g (GI+GII) at 10 km	Shallow estuary	Campos et al. (2015)
New Zealand (Dunedin-mouth of Andersons Bay)	Mussels	Reductions of 1 $\log_{10}$ (GI) and 1.3 $\log_{10}$ (GII) over 10 km. Mean concentrations of 15 copies/g (GI) and 34 copies/g (GII) at 10 km	Open coast environment	Greening and Lewis (2007)

simulate effects from all discharges impacting a given shellfish growing area. When planning a dye study, an initial concept of potential discharge effects on the growing area should be developed prior to the dye release. It is also important that the volumes of dye (or dye mixture) released in the environment result in concentrations below toxic thresholds for marine organisms. Field et al. (1995) provide reference toxicological information on chemical fluorescent dyes.

In recent years, there has been significant progress in identifying alternative viral indicators, raising the possibility of incorporating these indicators in routine monitoring of wastewater discharges and shellfish growing areas. Some of these studies report comprehensive datasets on enteric viruses in wastewater effluents subject to different types of treatment (Pouillot et al. 2015) and groups of viruses considered more suitable for use in shellfish growing area management, such as somatic coliphages, F-specific RNA phages and phages specific to human sources of faecal contamination, such as those infecting certain *Bacteroides* strains (Gyawali et al. 2021, da Silva et al. 2022). Reference concentrations in effluents and treatment removal rates of FIB (Kay et al. 2008) and NoV (Katayama and Vinjé 2017) are also available in the literature and can be used to estimate contaminant loading in the absence of site-specific data. However, significant data gaps exist for pathogen concentrations and viral indicators in certain types of discharges (CSOs, discharges from animal processing plants) and areas with large concentrations of birds and marine mammals.

A number of studies have reported data on the geographical abundance and distribution of NoV in shellfish growing areas (Table 1). In New Zealand, a gradient of NoV was observed in oysters as distance from the STW outfall. Total levels of NoV (GI + GII) were about 1,000 copies/g adjacent to the outfall and reduced to 130 copies/g at 10 km and to 100 copies/g at 24 km from the outfall (Greening

2007). In New South Wales (Australia), samples of Sydney rock oysters (*Saccostrea glomerata*) collected from sites along a river impacted by a sewage spill from a SO were positive for NoV GII up to 8.2 km downstream from the sewage source (Brake et al. 2018). These studies demonstrate the large areas of NoV contamination in coastal environments with very different characteristics.

2.2 *Monitoring and classification of shellfish production areas*

Monitoring of FIB in water/shellfish is an essential element of shellfish sanitation programmes worldwide (FAO and WHO 2020, 2021a). Concentrations of FIB provide an indication of the risk of growing area contamination with pathogens and helps determine the type of post-harvest treatment to which the shellfish must be subjected before marketing for human consumption. Results of faecal indicator monitoring are used to determine a classification category to each growing area against prescribed standards. The classification determines if the growing area is suitable for harvesting for direct human consumption, if additional post-harvest purification treatments are required to ensure that the shellfish meets quality and safety standards for human consumption or if the area is unsuitable for growing or harvesting shellfish for this purpose (FAO and WHO 2020).

It is standard good practice to review the sanitary survey and the classification status to ensure that the data reflect the level of risk at the time of the classification compliance assessment. The classification should be based on the analysis of a time series of monitoring results because concentrations of FIB in the environment vary largely in space and time as determined by the influence of environmental factors liketides, rainfall, wind strength and direction, bathymetry and water movements, etc. (Campos et al. 2013). In the UK, mean FIB concentrations in nearshore waters were found to vary $> 1 \log_{10}$ every day during base-flow conditions in nearby rivers, occasionally exceeding $2 \log_{10}$ (Wyer et al. 2018). Considering this large variability, the use of low numbers of monitoring results in the compliance assessment may result in large fluctuations in the classification status of a growing area which may not correspond to actual changes in the risk of contamination with pathogens. Consequently, it is important that the dataset used for growing area classification is sufficiently large to reduce the effects of environmental factors and statistically robust to detect any seasonal/annual variations in the classification status of the growing area. It is also important that all the information relevant to the classification status is readily accessible to the industry and consumers. Information on the species sampled, the location of sampling points for monitoring, the frequency of sampling and the methods used for analysis of monitoring data is compiled in a sampling plan which may be included in the sanitary survey report (Cefas 2018).

3. Additional control measures

3.1 *Buffer zones/exclusion zones around pollution sources*

Monitoring and classification of shellfish growing areas based on testing of FIB in water/shellfish has been effective in protecting consumers from exposure to bacterial pathogens. However, it is well established that enteric viruses have different survival characteristics to indicator bacteria in aquatic environments and therefore samples of shellfish with low concentrations of indicator bacteria (i.e., compliant with the standards for direct human consumption) cannot be guaranteed to be free from enteric viruses (Romalde et al. 2002, Campos and Lees 2014). Consequently, additional control measures are needed to mitigate viral risks, particularly in urbanised areas with large numbers of point-source inputs of faecal contamination. Shellfish safety authorities may designate buffer zones/exclusion zones around WWTP discharges, storm overflows, pump station overflows, septic tank discharges and/or areas with high concentrations of boats (e.g., marinas, ports) or mouths of rivers and streams (FAO and WHO 2021). Buffer zones may not be required if the sanitary survey indicates that the risk of contamination with pathogens is low.

Buffer zones around WWTP discharges and other sources of contamination have been implemented in many countries (e.g., USA, Canada, Italy, The Netherlands) (reviewed by Fitzgerald 2015). In the USA, the National Shellfish Sanitation Program (NSSP) contains requirements for buffer zones based on a theoretical calculation of sewage dilution required to meet a bacteriological standard of 14 faecal coliforms/100 ml in the growing water (USFDA and ISSC 2019). The calculations are based on worst-case microbiological loadings (untreated effluents for WWTP discharges) and the buffer zone must incorporate a minimum dilution of 1,000:1 of estuarine water to treated effluent (USFDA and ISSC 2019). Waters with dilutions less than 1,000:1 are classified as "prohibited" to shellfish harvesting. In the absence of site-specific monitoring data, the NSSP recommends that a concentration of 1.4×10^6 faecal organisms/100 ml of water be assumed for untreated wastewater and that a 100,000:1 dilution be considered to meet the target 14 faecal coliforms or *E. coli*/100 ml in the receiving waters (FAO and WHO 2021). In Europe, the Food Hygiene Regulations do not contain requirements for buffer zones although some EU Member States have implemented some form of zone-based controls based on geographical proximity to pollution sources as required by national legislation (Cefas 2013). The Guide to Good Practice on Microbiological Monitoring of Bivalve Mollusc Harvesting Areas includes recommendations on buffer zones around sewage outfalls, harbours and marinas (Cefas 2018). The approaches to defining buffer zones vary between and within Member States. Most apply fixed zoning relative to distance from the contamination source(s) (Cefas 2013a). In Italy, exclusion zones have been implemented in Marche, Veneto, and Campania regions. In the latter, protocols determine that, in the absence of information on the circulation of pollutants, water bodies located within a 500 m radius of each contaminating source should not be classified for the commercial production of shellfish (Cefas 2013a). In the Netherlands, exclusion zones are determined by national legislation and have been determined for areas adjacent to wastewater discharges and in ports and river mouths, with radii ranging from 100 m to 1,500 m (Cefas 2013a).

A review of the international literature indicates that the basic information requirements for determining a buffer zone are:

- location, volume, flow, periodicity, and microbiological quality of the wastewater discharge
- concentration and decay rate of the microbiological contaminants of interest

- characteristics of the receiving environment such as tides, bathymetry, water circulation, dilution rate
- location of the shellfish resources in relation to the pollution source(s)
- time of wastewater transport to the shellfish growing area.

The rate at which microbiological contaminants are inactivated in surface waters is very dependent upon temperature and UV irradiance (Campos et al. 2013). Decay rates as a function of sunlight, temperature and water matrix have been published for a range of human-associated indicators and pathogens (e.g., *E. coli*, norovirus, *Salmonella*, *Campylobacter*, *E. coli* O157:H7, *Giardia* and *Cryptosporidium* (Boehm et al. 2018). Pathogen survival is shorter in the tropics than in temperate climates (days-weeks), and can extend to months in the winter in colder latitudes. Pathogens can persist to remain infectious for some time in faecally-contaminated environments, particularly in cold climates, where they can be re-mobilised from land during flood events and snowmelt. Other factors, such as sunlight, can reduce pathogen concentrations in shallow waters. Temperature, association with solids and microbiological flora are the main factors affecting the persistence of enteric viruses in aquatic environments (Bosch et al. 2006, USEPA 2015). FIB are less persistent (hours-days) in the environment than enteric viruses (days-weeks) because of their smaller size, greater stability over a wide range of temperature and pH values, greater resistance to chemical agents, and high affinity to sediments.

Different approaches exist to assess the circulation of contaminants in discharge receiving environments, ranging from qualitative assessments of bathymetry and current velocities to quantitative methods such as simple dilution-based models, dye tracing and drogue tracking studies, or more comprehensive hydrodynamic, water quality and particle tracking models. Initially, each pollution source should be assessed individually to identify which source(s) has the greatest effect on the growing areas. In coastal areas with high numbers of small volume discharges, it may be appropriate to consider discharge agglomerations and determine their combined effects on the microbiological status of the growing area.

3.2 *Active management of harvesting activity based on environmental triggers*

Active management systems use surrogate variables (e.g., rainfall, wastewater discharge volumes, river flows, salinity, water temperature) to predict microbiological contamination in shellfish growing areas and to inform management of harvesting activity, including closure/re-opening of harvesting activity following extreme weather events, wastewater spills or notification of illness cases. In Canada, shellfish safety authorities are required to place shellfish growing areas in the closed status following significant weather events, flooding and wastewater spills (CFIA et al. 2022). The affected growing area remains closed for at least seven days until sampling of water and shellfish indicates that the microbiological status of the growing area has returned to normal, i.e., complies with its classification criteria. If sampling is not undertaken, the growing area remains in the closed status over at least 21 days (CFIA et al. 2022). If the closure is caused by the presence of pathogens in the growing area, the area remains closed for at least 30 days, until test results are deemed acceptable (CFIA et al. 2022). In Australia, shellfish growing areas affected by discharges of untreated/partially treated wastewater must be closed over at least 21 days since the end of the contamination event or when shellfish samples, collected from representative locations in the growing area (no sooner than seven days after the contamination has ceased), contain male-specific coliphage levels below background levels or 50 pfu/100 g (ASQAP 2019). In France, the harvesting of shellfish from areas linked to norovirus outbreaks must be prohibited as soon as the first case of illness occurs and investigations demonstrate a link to a contaminated growing area. The growing area can be reopened once shellfish samples are negative for norovirus or, alternatively, if a rainfall threshold is not exceeded during 28 days following the outbreak, or there are no further notifications of WWTP failures and the results of *E. coli* monitoring return to normal (Ministry of Agriculture, Food Processing and Forestry 2012). In Aotearoa New Zealand, there are similar closure requirements for emergency discharges due to extreme rainfall, river flows, or broken sewer pipe or storm/flood event (MPI 2018). The closure criteria are specified in the growing area management plan. If the growing area has been implicated in an illness outbreak of viral aetiology, the shellfish authority must review the classification monitoring data, investigate the source of the contamination, and keep the area closed for 28 days until the event that caused the closure no longer exists and the pathogen is no longer present (MPI 2018). Shellfish growing area closures generate considerable social and economic impacts on the industry. In the Machias Bay region of Maine (USA), Evans et al. (2016) found that closures due to pollution contributed to a loss of $3.6 million in forgone revenue (27.4% of total revenue) over the period 2001–2009. Closures linked to CSO discharges generated most of these losses ($2 million).

Active management systems are increasingly incorporating artificial intelligence (artificial neural networks) (Wang and Deng 2019) and sensor technology to provide near real-time information on microbiological loadings from pollution sources and the fate and transport of microbiological contaminants within catchments and coastal and marine environments. Sensor technology supplies data from monitoring buoys (Schmidt et al. 2018, 2018a), rain and river gauges (Zimmer-Faust et al. 2018), remote sensing (Chenar and Deng 2018, 2021) and shellfish physiological and behavioural sensors (biotags) (Andrewartha et al. 2015). The rationale for active management systems may be provided in sanitary survey reports with support from statistical analyses of microbiological results and surrogate environmental parameters. In New Zealand, a catchment and coastal hydrodynamic modelling tool has been developed and is now operational. It supports shellfish farm management in response to changing coastal conditions and microbiological contamination risk (Figure 1). A coastal monitoring platform placed near shellfish farms supported hydrodynamic model calibration and validation and is now a key operational driver of the model, providing temperature, salinity, turbidity and chlorophyll data.

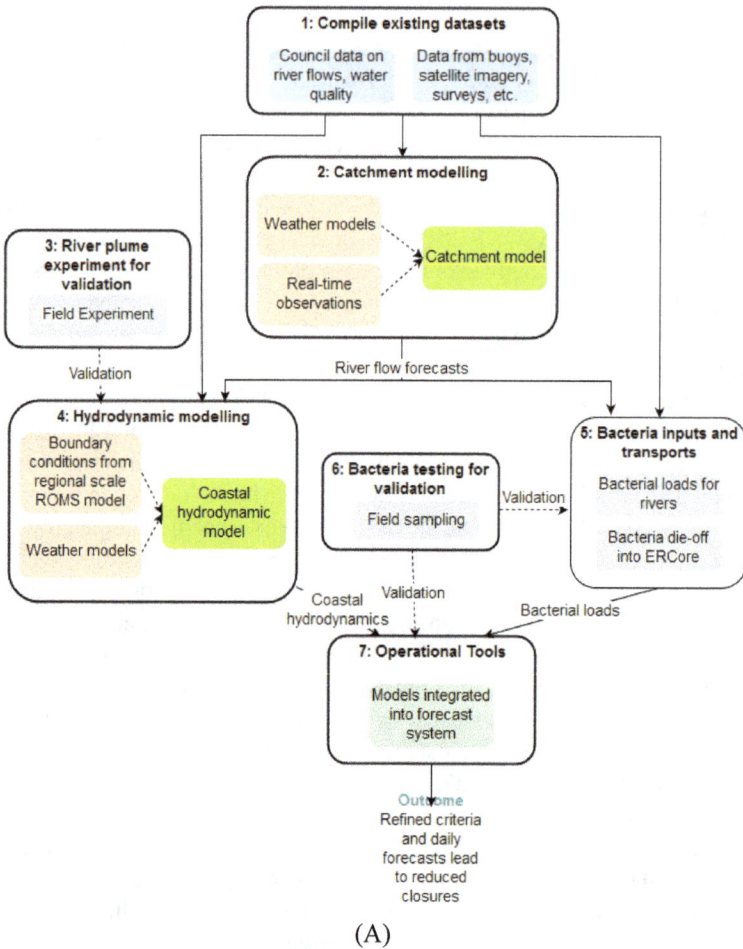

(A)

(B)

Figure 1. Components of a catchment-to-coast forecasting tool and monitoring buoy (B) used for microbiological risk management of shellfish growing areas (A) in Aotearoa New Zealand.

3.3 Pathogen monitoring and criteria for growing area management

Pathogen monitoring has been undertaken in growing areas with a history of outbreaks (Rupnik et al. 2018) and when the sanitary survey indicates that there is a potential high health risk during certain periods of the year and/or concentrations of FIB are elevated well above the normal range potentially causing periodic closures of the growing area. Baseline data on NoV in shellfish growing areas now exist for a number of countries:

- UK: 76% in oysters (*C. gigas, O. edulis*) (Lowther et al. 2012)
- Europe (12 Member States and Norway; mean apparent prevalence): 34% in oysters (*C. gigas, O. edulis, C. angulata*) (EFSA 2019)
- France: 22% in oysters, mussels and common cockles (Rincé et al. 2018)
- USA: 20% in oysters (Costantini et al. 2006)
- Korea: 2–12% in oysters (Moon et al. 2011)
- Australia: < 2% in oysters (*C. gigas, S. glomerata*) (Torok et al. 2018)

These studies show marked differences in NoV prevalence in different geographies. However, comparisons between countries are not appropriate because of the differences in method performance,sampling and data analysis approaches. While these data are very useful to support risk assessments, there is little information on species-specific rates of viral bioaccumulation and clearance from the shellfish, particularly during short-term contamination events. Environmental studies have found that NoV concentrations in shellfish can reduce when the shellfish are transferred from contaminated to clean waters (relaying) (McLeod et al. 2017). In Ireland, batches of oysters transferred to clean waters over 17 days had lower (0.8 $\log_{10}$) concentrations of NoV than oysters harvested from their commercial growing area (Doré et al. 2010). A further reduction of 0.6 $\log_{10}$ was achieved when the re-laid oysters were placed in purification tanks at 17°C over 4 days. McLeod et al. (2017) suggested that a relay period of about 4 weeks may be sufficient to reduce NoV GI and GII to background levels in Pacific oysters and viral infectivity would also reduce during this time.

In the European Union, the EFSA Panel on Biological Hazards has recommended the introduction of microbiological criteria for NoV in shellfish in the food hygiene regulations, unless batches are labelled "to be cooked before consumption" (EFSA 2012). The Panel has further recommended refinement of the regulatory standards and monitoring approaches to improve public health protection (EFSA 2012). Scope for these recommendations is given in Article 27 of Regulation (EC) No. 2073/2005: "In particular, criteria for pathogenic viruses in live bivalve molluscs should be established when the analytical methods are developed sufficiently." (European Communities 2005, p. 4). Although the implementation of NoV standard(s) would reduce the number of contaminated batches placed on the market, the available evidence is insufficient to estimate the impact of any potential standards on consumers' exposure to NoV. The Food Safety Authority of Ireland has recommended that oysters from a class A production area implicated in NoV outbreaks can only be sold for human consumption if NoV concentrations in the oysters are < 200 copies/g based on results of two consecutive samples taken at least 24 hours apart. For oysters from class B areas, NoV concentrations in the oysters following post-harvest treatment (e.g., depuration at increased temperature) must be < 200 copies/g (Food Safety Authority of Ireland 2013). The use of male specific coliphage as a viral indicator for growing area closure/reopening has been advocated. This has been supported by much research over the past 20 years with criteria incorporated into a number of shellfish sanitation programmes and recommended in international good practice guidance (FAO and WHO 2021, Pouillot et al. 2021). While testing of male specific coliphage in conjunction with FIB/NoV is useful to determine viral risks associated with reduced wastewater treatment efficiency, wastewater spill events, and viral outbreaks, it must be acknowledged that this testing is not appropriate in all circumstances and requires consideration of other environmental data and a weight of evidence approach to be effective (Hay et al. 2013, Hodgson et al. 2015).

4. Gaps in evidence and future perspectives

The microbiological quality of shellfish growing areas is highly variable in space and time and results from a complex interaction of multiple environmental factors. The risk management framework incorporates two fundamental components (sanitary surveys and microbiological monitoring and classification of growing areas) which serve multiple functions. Sanitary surveys produce qualitative or quantitative information on the sources of faecal contamination affecting the growing areas and the potential effects of these sources on the microbiological status of the areas. On the other hand, monitoring helps to determine if shellfish harvesting for human consumption can take place in the growing area and the controls required to maintain the microbiological risk at low or acceptable levels. The outcomes of the sanitary survey may also determine the contents of sampling plans for monitoring, namely the appropriate number and location of sampling sites and frequency of sampling.

Analysis of sanitary survey reports from different countries indicated that, in some cases, there is lack of detail in key elements of the surveys (e.g., fate and transport of microbiological contaminants; microbiological loading from key freshwater inputs and intermittent discharges) and this means that fully quantitative risk assessments cannot be made. Furthermore, the rationale for selecting sampling sites is not always sufficiently discussed in the reports. This lack of detail could be due to the level of specification given in the legislation, lack of resources, and/or poor technical application of survey methodologies.

An important factor that affects compliance with classification standards is the species-specific rate of bioaccumulation and clearance of microbiological contaminants. Comparative data on average concentrations of microbiological contaminants in different species of shellfish exists only for a few geographical locations. In the UK, Younger and Reese (2013) reported that, on average, clams, cockles and mussels accumulate *E. coli* to higher concentrations than oysters. These differences occur because of morphological and physiological characteristics of the different species. Similar data are not available for viral contaminants. Similar results found in France indicated that median *E. coli* concentrations in clams (*Tapes* spp.) are 4x higher than in oysters (*C. gigas*) (Amouroux and Soudant 2011). Further comparative data on the rates of accumulation of FIB and enteric viruses are required to identify the most appropriate species for sampling in growing areas where multiple species are harvested.

Monitoring of large shellfish growing areas based on a reduced number sampling sites or samples taken monthly (or less frequently) is unlikely to provide a good measure of microbiological risk. In many monitoring programmes, sampling is mostly undertaken during wet weather and/or certain states of the tide. Limiting sampling to these conditions will likely bias the assessment of the microbiological status of the growing areas. Furthermore, many monitoring programmes rely on testing of samples of water/shellfish taken infrequently (monthly is the most common frequency). Because traditional culture-based testing for FIB takes 18–24 hours to produce a result (normally 48 hours from sample receipt to result), consumers cannot be informed of the potential health risks in a timely manner in the event of unexpected contamination. Walker et al. (2020) developed an alternative, rapid method in which the final *E. coli* confirmation step in the Most Probable Number technique (TBX culture) was replaced by presence/absence real-time PCR (qPCR). This method would require full validation before it can be used with confidence by official control laboratories (Walker et al. 2020).

Another important knowledge gap relates to natural environmental reservoirs of FIB or "naturalized" populations in stream sediments and vegetation (Devane et al. 2020). These reservoirs are areas where faecal bacteria are stored during low flows and survive for longer periods. The microbes can be remobilised back into the water column following rainfall or other disturbance events. It has been demonstrated that while *E. coli* can survive for months in river sediments and other in-channel reservoirs, some pathogens do not survive well in these environments. In nutrient-rich sediments, *E. coli* may actually grow rather than simply survive. Our understanding of these 'naturalized' populations of faecal bacteria needs to improve. The presence of naturalized indicator bacteria is likely to confound the relationships between FIB and pathogens and therefore health risk assessments during periods of non-compliance with regulatory standards.

While it is still common practice to use FIB to infer risk of illness from pathogens, many studies have highlighted differences in the potential/actual health risks from human and non-human sources of contamination. These differences are associated with the type of source and the abundance and virulence of pathogens associated with each individual source. Whole genome sequencing (WGS) techniques determine the complete DNA sequence of an organism's genome at a single time and therefore provide high resolution microbiological subtyping. When used as part of microbiological monitoring programmes, WGS techniques can significantly increase the speed at which pathogenic hazards are detected leading to more targeted and effective risk management interventions. WGS can help inform selection of pathogenic hazards for monitoring (particularly as part of outbreak investigations), confirm the presence of hazards of health concern not detected through routine monitoring, and pollution source attribution/apportionment (Lozano-León et al. 2021). However, the health significance of detecting genomic material in the environment is unknown, i.e., genomic material may be present even when the microorganism is no longer infectious or inactive. As WGS is increasingly implemented worldwide, there is a need to develop guidance on interpretation of results and undertake method validation to achieve greater confidence in results.

The main viruses of concern in relation to shellfish safety are NoV and Hepatitis A. It is likely that viral standards will be included in the regulatory framework in the next few years. Before this can happen, further research is needed to identify suitable viruses and methods that are suitably sensitive and can be applied with confidence by official control laboratories. Further studies comparing pathogen diversity in wastewater with that in artificially/naturally contaminated shellfish such as that reported by Strubbia et al. (2019) are needed. There have also been considerable developments in rapid test kits for a range of viral indicators. These tools need to be sensitive, portable, fast and it is important that their results correlate well with traditional laboratory methods. Method performance comparison studies have been undertaken internationally to provide a better understanding of equivalence between methods and greater confidence in test results.

The general public has limited understanding of the highly variable nature of microbiological contamination in the environment. There is the perception that the shellfish can be deemed either safe/unsafe based on a single or few test results. In reality, the risk of microbiological contamination in growing areas as indicated by FIB changes considerably over short spatial and temporal scales. Much is known about the sources, transport and fate of FIB in temperate climates, but further studies on the factors controlling the environmental persistence of FIBs in tropical environments are needed (Rochelle-Newall et al. 2015). Most of the data on the persistence and dispersion of FIB comes from studies conducted in either temperate systems or in tropical systems in developed countries. Little or no data are available on these parameters in developing nations lacking infrastructure for collection and treatment of municipal and industrial wastewater.

Near real-time monitoring of microbiological water and shellfish quality provides the shellfish industry and consumers with detailed information on spatio-temporal variation of microbiological contaminants in growing areas (de Souza et al. 2018). A number of forecasting tools have been developed to support a range of applications, including identification of suitable sites for shellfish farming, identification of sampling sites and testing of criteria for closure/reopening of harvesting activity. Forecasting tools include simple decision trees and notification protocols, statistical regression models, rainfall-based triggers, deterministic models, and artificial intelligence methods. Many of these models provide same-day predictions, including on weekends, when the shellfish growing areas are not being monitored. This is vital information for the industry. However, not all growing areas are suitable for use of predictive modelling. For instance, those that rarely exceed the classification limit may not benefit from these tools.

Irrespective of the type of model used, these tools require large monitoring datasets for development and are associated with a degree of uncertainty which can be reduced over time through validation. Perhaps the greatest limitation of these tools is their reduced ability to simulate viral contamination. This stems from a poor understanding of the relationships between viral contamination in water and shellfish which has led, in some cases, to poor model parameterization. This is a fundamental component of the risk management process which requires further research to generate empirical data to calibrate the models. Key to this is the definition of the minimum number of samples and sampling strategy for model development and clear guidance on which circumstances model predictions should be used instead of sampling during periods of non-compliance. With greater acceptance and implementation of these modelling tools as part of regulatory monitoring programmes, there is a need to provide further context around predictions, including measures of risk alongside microbiological concentrations and how they relate to traditional regulatory monitoring based on 'spot' sampling. Further guidance is also needed on how to maintain the models and their transferability to other shellfish growing areas. Despite the high cost normally associated with hydrodynamic modelling, model implementation can contribute important information to inform decisions on harvesting closures and, in some cases, represent significant cost savings to the industry.

Microbiological monitoring programmes require significant resources and there is plenty of opportunity to improve monitoring plans to reflect the latest scientific and technological developments. Faecal contamination of rivers and coastal waters is a global problem that can be reversed (Landrigan et al. 2020). A vision for improved management of microbiological risks in shellfish growing areas is presented in Figure 2. The vision proposes the following key components:

- Implementation of cost-effective methods for identifying and quantifying the microbiological loading from individual sources.
- Robust data management and statistical analysis of environmental data collected in monitoring programmes.
- Provision of information on environmental risk factors and health risk in a timely manner through near-real time monitoring and/or forecasting of microbiological contamination.
- Improved characterization of the relationships between FIB and pathogens in shellfish growing waters.
- Communication of sanitary survey results and classification status of growing areas via community online portals.
- Citizen-science initiatives to engage shellfish consumers in sanitary surveys and water quality investigations (catchment walkover surveys, sample collection, data reporting, etc.).
- Regular dialogue between shellfish farmers, regulators, policy-makers and consumer organizations.
- Regular updating of protocols and procedures to ensure that they reflect the latest scientific developments.

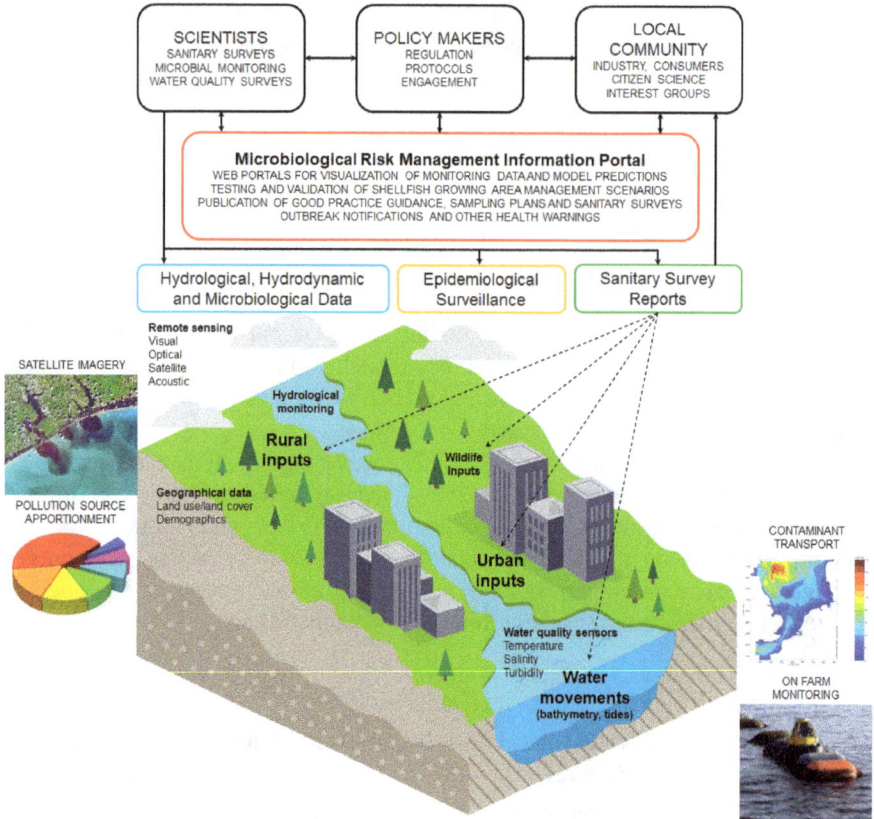

Figure 2. A vision for microbiological monitoring of shellfish growing areas.

Abbreviations

QMRA	quantitative microbial risk Assessment
MST	Microbial Source Tracking
NoV	Norovirus
PCR	Polymerase Chain Reaction
WGS	Whole Genome Sequencing
WWTP	Wastewater Treatment Plant

References

Ahmed, S.M., Lopman, B.A. and Levy, K. 2013. A systematic review and meta-analysis of the global seasonality of norovirus. PLoS ONE 8(10): e75922.

Amouroux, I. and Soudant, D. 2011. Comparison of microbiological contamination level between different species of shellfish. Presented at the 8th International Conference on Molluscan Shellfish Safety, ICMSS11, Charlottetown, Canada, June 12–17 2011. https://archimer.ifremer.fr/doc/00176/28731/.

Andrewartha, S.J., Elliott, N.G., McCulloch, J.W. and Frappell, P.B. 2015. Aquaculture sentinels: smart-farming with biosensor equipped stock. Journal of Aquaculture Research & Development 7: 1.

Ao, Y. and Goblick, G.N. 2016. Application of hydrodynamic modelling to predict viral impacts from wastewater treatment plant discharges adjacent to shellfish growing areas.

https://19january2017snapshot.epa.gov/sites/production/files/2016-06/documents/application-hydrodynamic-modeling.pdf.

ASQAP. 2019. Australian shellfish quality assurance program. Operations manual. Version 5. Australian Shellfish Quality Assurance Advisory Committee. hhttps://www.safefish.com.au.

Bellou, M., Kokkinos, P. and Vantarakis, A. 2013. Shellfish-borne viral outbreaks: a systematic review. Food and Environmental Virology 5(1): 13–23.

Boehm, A.B., Graham, K.E. and Jennings, W.C. 2018. Can we swim yet? Systematic review, meta-analysis, and risk assessment of aging sewage in surface waters. Environmental Science and Technology 52(17): 9634–9645.

Borchardt, M.A., Bradbury, K.R., Alexander Jr, E.C., Kolberg, R.J., Archer, J.R., Braatz, L.A. et al. 2011. Norovirus outbreak caused by a new septic system in a Dolomite aquifer. Ground Water 49: 85–97.

Bosch, A., Pintó, R.M. and Abad, F.X. 2006. Survival and transport of enteric viruses in the environment. *In*: Goyal, S.M. (ed.). Viruses in Foods. Food Microbiology and Food Safety. Springer, Boston.

Brake, F., Kiermeier, A., Ross, T., Holds, G., Landinez, L. and McLeod, C. 2018. Spatial and temporal distribution of norovirus and *E. coli* in Sydney Rock Oysters following a sewage overflow into an estuary. Food and Environmental Virology 10: 7–15.

Brown, A.R., Webber, J., Zonneveld, S., Carless, D., Jackson, B., Artioli, Y. et al. 2020. Stakeholder perspectives on the importance of water quality and other constraints for sustainable mariculture. Environmental Science & Policy 114: 506–518.

Butt, A.A., Aldridge, K.E. and Sanders, C.V. 2004. Infections related to the ingestion of seafood. Part I: viral and bacterial infections. The Lancet Infection Diseases 4(4): P201–212.

Campos, C.J.A. and Cachola, R.A. 2007. Faecal coliforms in bivalve harvesting areas of the Alvor Lagoon (southern Portugal): influence of seasonal variability and urban development. Environmental Monitoring and Assessment 133(1–3): 31–41.

Campos, C.J.A., Hargin, K., Kershaw, S., Lee, R.J. and Morgan, O.C. 2011. Rainfall and river flows are predictors for β-glucuronidase positive *Escherichia coli* accumulation in mussels and Pacific oysters from the Dart Estuary (England). Journal of Water and Health 9(2): 368–381.

Campos, C.J.A., Kershaw, S. and Lee, R.J. 2013. Environmental influences on faecal indicator organisms in coastal waters and their accumulation in bivalve shellfish. Estuaries and Coasts 36: 834–853.

Campos, C.J.A. and Lees, D.N. 2014. Environmental transmission of human noroviruses in shellfish waters. Applied and Environmental Microbiology 80(12): 3552–3561.

Campos, C.J.A., Avant, J., Gustar, N., Lowther, J., Powell, A., Stockley, L. et al. 2015. Fate of human noroviruses in shellfish and water impacted by frequent sewage pollution events. Environmental Science & Technology 49(14): 8377–8385.

Campos, C.J.A., Avant, J., Lowther, J., Till, D. and Lees, D.N. 2016. Human norovirus in untreated sewage and effluents from primary, secondary and tertiary treatment processes. Water Research 103: 224–232.

Campos, C.J.A., Goblick, G., Lee, R., Wittamore, K. and Lees, D.N. 2017. Determining the zone of impact of norovirus contamination in shellfish production areas through microbiological monitoring and hydrographic analysis. Water Research 124: 556–565.

Campos, C.J.A., Kershaw, S., Morgan, O.C. and Lees, D.N. 2017. Risk factors for norovirus contamination of shellfish water catchments in England and Wales. International Journal of Food Microbiology 241: 318–324.

Campos, C.J.A., Morrisey, D.J. and Barter, P. 2022. Principles and technical application of mixing zones for wastewater discharges to freshwater and marine environments. Water 14: 1201.

Cardemil, C.V., Parashar, U.D. and Hall, A.J. 2017. Norovirus infection in older adults. Epidemiology, risk factors, and opportunities for prevention and control. Infectious Diseases Clin. North Am. 31(4): 839–870.

Cefas. 2013. Discussion paper on live bivalve molluscs and human enteric virus contamination: options for improving risk management in EU food hygiene package. Cefas Weymouth.

Cefas. 2013a. Report of the 12th workshop of NRLs for monitoring bacteriological and viral contamination of bivalve molluscs. Prohibition zones – EU situation. https://eurlcefas.org/media/13832/prohibtion_zones_workshop_of-nrls_for_bacteriological_and_viral_contamination_of_bivalve_molluscs_final.pdf.

Cefas. 2018. Microbiological monitoring of bivalve mollusc harvesting areas. Guide to good practice: technical application. EU Working Group on the Microbiological Monitoring of Bivalve Mollusc Harvesting Areas.

CFIA, DFO and ECCC. 2022. Canadian shellfish sanitation program manual. https://inspection.canada. ca/food-safety-for-industry/food-specific-requirements-and-guidance/fish/canadian-shellfish-sanitation-program/eng/1527251566006/1527251566942?chap=0.

Chenar, S.S. and Deng, Z. 2018. Development of artificial intelligence approach to forecasting oyster norovirus outbreaks along Gulf of Mexico coast. Environment International 111: 212–223.

Chenar, S.S. and Deng, Z. 2021. Hybrid modeling and prediction of oyster norovirus outbreaks. Journal of Water and Health 19(2): 254–266.

Cho, K.H., Pachepsky, Y.A., Oliver, D.M., Muirhead, R.W., Park, Y., Quilliam, R.S. et al. 2016. Modeling fate and transport of fecally-derived microorganisms at the watershed scale: state of the science and future perspectives. Water Research 100: 38–56.

Colaiuda, V., di Giacinto, F., Lombardi, A., Ippoliti, C., Giansante, C., Latini, M. et al. 2021. Evaluating the impact of hydrometeorological conditions on E. coli concentration in farmed mussels and clams: experience in central Italy. Journal of Water and Health 19.3: 512–533.

Costantini, V., Loisy, F., Joens, L., Le Guyader, F.S. and Saif, L.J. 2006. Human and animal enteric caliciviruses in oysters from different coastal regions of the United States. Applied and Environmental Microbiology 72(3): 1800–1809.

da Silva, D.T.G., Ebdon, J., Dancer, D., Baker-Austin, C. and Taylor, H. 2022. A longitudinal study of bacteriophages as indicators of norovirus contamination of mussels (Mytilus edulis) and their overlying waters. Pollutants 2: 66–81.

de Souza, R.V., Campos, C.J.A., Garbossa, L.P.H. and Seiffert, W.Q. 2018. Developing, cross-validating and applying regression models to predict the concentrations of faecal indicator organisms in coastal waters under different environmental scenarios. Science of the Total Environment 630: 20–31.

Devane, M.L., Moriarty, E., Weaver, L., Cookson, A. and Gilpin, B. 2020. Fecal indicator bacteria from environmental sources; strategies for identification to improve water quality monitoring. Water Research 185: 116204.

Dodgson, R.W. 1928. Report on mussel purification. Ministry of Agriculture and Fisheries. Investigations, Series II, Vol. 10. HMSO, London.

Doré, B., Keaveney, S., Flannery, J. and Rajko-Nenow, P. 2010. Management of health risk associated with oysters harvested from a norovirus contaminated area, Ireland, February–March 2010. EuroSurveillance 15(19): 1–5.

EFSA. 2012. Scientific opinion on norovirus (NoV) in oysters: methods, limits and control options. EFSA Journal 10: 2500.

EFSA. 2019. Analysis of the European baseline survey of norovirus in oysters. EFSA Journal 17(7): 5762.

European Communities. 2005. Commission Regulation (EC) No. 2073/2005 of 15 November 2005 on microbiological criteria for foodstuffs. Official Journal of the European Communities L338, 22.12.2005: 1–26.

Evans, K.S., Athearn, K., Chen, X., Bell, K.P. and Johnson, T. 2016. Measuring the impact of pollution closures on commercial shellfish harvest: the case of soft-shell clams in Machias Bay, Maine. Ocean & Coastal Management 130: 196–204.

FAO, WHO. 2020. Code of Practice for Fish and Fishery Products. Rome.

FAO, WHO. 2021. Microbiological risk assessment: guidance for food. Microbiological Risk Assessment Series No. 36. Rome.

FAO, WHO. 2021a. Technical guidance for the development of the growing area aspects of Bivalve Mollusc Sanitation Programmes. Second edition. Food Safety and Quality Series No. 5.

Field, M.S., Wilhelm, R.G., Quinlan, J.F. and Aley, T.J. 1995. An assessment of the potential adverse properties of fluorescent tracer dyes used for groundwater tracing. Environmental Monitoring and Assessment 38: 75–96.

Fitzgerald, A. 2015. Review of approaches for establishing exclusion zones for shellfish harvesting around sewage discharge points. Desk study to inform consideration of the possible introduction of exclusion zones as a control for norovirus in oysters. Aquatic Water Services Ltd and Aquafish Solutions Ltd report No FS513404 to the Food Standards Agency. https://www.food.gov.uk/sites/default/files/media/document/FS513404%20-%20FINAL.pdf.

Food Safety Authority of Ireland. 2013. Risk management of norovirus in oysters. Opinion by the Food Safety Authority of Ireland Scientific Committee, December 2013. https://www.fsai.ie/publications_norovirus_opinion/.

Fumian, T.M., Fioretti, J.M., Lun, J.H., dos Santos, I.A.L., Whote, P.A. and Miagostovich, M.P. 2019. Detection of norovirus epidemic genotypes in raw sewage using next generation sequencing. Environment International 123: 282–291.

Galtsoff, P.S. 1964. The American Oyster, *Crassostrea virginica*, Gmelin. Fisheries Bulletin of the US Fish and Wildlife Service 64. Washington DC: US Government Printing Office.

Garbossa, L.H.P., Souza, R.V., Campos, C.J.A., Vanz, A., Vianna, L.F.N. and Rupp, G.S. 2017. Thermotolerant coliform loadings to coastal areas of Santa Catarina (Brazil) evidence the effect of growing urbanisation and insufficient provision of sewerage infrastructure. Environmental Monitoring and Assessment 189(1): 27.

García-García, L.M., Campos, C.J.A., Kershaw, S., Younger, A. and Bacon, J. 2021. Scenarios of intermittent *E. coli* contamination from sewer overflows to shellfish growing waters: the Dart Estuary case study. Marine Pollution Bulletin 167: 112332.

Garreis, M.J. 1994. Sanitary surveys of growing waters. pp. 289–330. *In*: Hackney, C.R. and Pierson, M.D. (eds.). Environmental Indicators and Shellfish Safety. New York: Chapman and Hall.

Goblick, G.N., Anbarchian, J.M., Woods, J., Burkhardt III,W. and Calci, K. 2011. Evaluating the dilution of wastewater treatment plant effluent and viral impacts on shellfish growing areas in Mobile Bay, Alabama. Journal of Shellfish Research 30(3): 979–987.

Goblick, G.N., Ao, Y., Anbarchian, J.M. and Calci, K.R. 2016. Determination of buildup and dilution of wastewater effluent in shellfish growing waters through a modified application of super-position. Marine Pollution Bulletin 115(1–2): 164–171.

Green, T.J., Walker, C.Y., Leduc, S., Michalchuk, T., McAllister, J., Roth, M. et al. 2022. Spatial and temporal pattern of norovirus dispersal in an oyster growing region in the Northwest Pacific. Viruses 14: 762.

Greening, G.E. 2007. NLRC shellfish safety following sewage spills. Envirolink report No. 162 NLRC20. https://envirolink.govt.nz/assets/Envirolink/162-nlrc20.pdf.

Greening, G.E. and Lewis, G.D. 2007. Virus prevalence in New Zealand shellfish. ESR Report No. 0659: FRST Programme C03X0301. Safeguarding Environmental Health and Market Access for NZ Foods. Objective 2: Virus Prevalence in Shellfish.

Guillon-Cottard, I., Augier, H., Console, J.J. and Esmieu, O. 1998. Study of microbiological pollution of a pleasure boat harbour using mussels as bioindicators. Marine Environmental Research 45(3): 239–247.

Gyawali, P. and Hewitt, J. 2020. Faecal contamination in bivalve molluscan shellfish: can the application of the microbial source tracking method minimise public health risks? Current Opinion in Environmental Science and Health 16: 14–21.

Gyawali, P., Devane, M., Scholes, P. and Hewitt, J. 2021. Application of crAssphage, F-RNA phage and pepper mild mottle virus as indicators of human faecal and norovirus contamination in shellfish. Science of the Total Environment 783: 146848.

Hall, A.J., Vinje, J., Lopman, B., Park, G.W., Yen, C., Gregoricus, N. et al. 2011. Updated norovirus outbreak management and disease prevention guidelines. Morbidity and Mortality Weekly Report 60(3): 1–15.

Harwood, V.J., Staley, C., Badgley, B.D., Borges, K. and Korajkic, A. 2014. Microbial source tracking markers for detection of fecal contamination in environmental waters: relationships between pathogens and human health outcomes. FEMS Microbiology Reviews 38(1): 1–40.

Hassard, F., Sharp, J.H., Taft, H., LeVay, L., Harris, J.P., McDonald, J.E. et al. 2017. Critical review on the public health impact of norovirus contamination in shellfish and the environment: a UK perspective. Food and Environmental Virology 9: 123–141.

Hata, A., Katayama, H., Koima, K., Sano, S., Kasuga, I., Kitajima, M. et al. 2014. Effects of rainfall events on the occurence and detection efficiency of viruses in river water impacted by combined sewer overflows. Science of the Total Environment 468-469: 757–763.

Hay, B., McCoubrey, D.J. and Zammit, A. 2013. Improving the management of the risk of human enteric viruses in shellfish at harvest. Case studies of oyster growing areas implicated in norovirus illness events. A Report by AquaBio Consultants Ltd, Dorothy-Jean & Associates Ltd and New South Wales Food Authority.

Hodgson, K., Torok, V., Jolley, J., Malhi, N. and Turnbull, A. 2015. Oysters Australia IPA: the use of FRNA bacteriophages for rapid re-opening of growing areas after sewage spills. A Report to the Fisheries Research and Development Corporation.

Hudson, D. 2011. Business case proposal for investment in the development of SPR technology for use by oyster industry stakeholders in the rapid identification and monitoring of norovirus (NoV), SGA Limited, Australia.

Iwai, M., Hasegawa, S., Obara, M., Nakamura, K., Horimoto, E., Takizawa et al. 2009. Continuous presence of noroviruses and sapoviruses in raw sewage reflects infections among inhabitants of Toyama, Japan (2006 to 2008). Applied and Environmental Microbiology 75: 1264–1270.

Iwamoto, M., Ayers, T., Mahon, B.E. and Swerdlow, D.L. 2010. Epidemiology of seafood-associated infections in the United States. Clinical Microbiology Reviews 23(2): 399–411.

Katayama, H., Haramoto, E., Oguma, K., Yamashita, H., Tajima, A., Nakajima, H. et al. 2008. One-year monthly quantitative survey of noroviruses, enteroviruses, and adenoviruses in wastewater collected from six plants in Japan. Water Research 42: 1441–1448.

Katayama, H. and Vinjé, J. 2017. Norovirus and other calicivirus. *In*: Rose, J.B. and Jiménez-Cisneros, B. (eds.). Water and Sanitation for the 21st Century: Health and Microbiological Aspects of Excreta and Wastewater Management (Global Water Pathogen Project). Meschke, J.S. and Girones, R. (eds.). Part 3: Specific Excreted Pathogens: Environmental and Epidemiology Aspects - Section 1: Viruses), Michigan State University, E. Lansing, MI, UNESCO. https://www.waterpathogens.org/book/norovirus-and-other-caliciviruses.

Kay, D., Crowther, J., Stapleton, C.M., Wyer, M.D., Fewtrell, L., Edwards, A. et al. 2008. Faecal indicator organism concentrations in sewage and treated effluents. Water Research 42: 442–454.

Kay, D., Anthony, Crowther, J., Chambers, B.J., Nicholson, F.A., Chadwick, D. et al. 2010. Microbial water pollution: a screening tool for initial catchment-scale assessment and source apportionment. Science of the Total Environment 408: 5649–5656.

Kershaw, S., Cook, A. and Campos, C. 2012. Sanitary surveys (England & Wales). Review of progress, processes and outcomes 2007–2011. Cefas report C3018 to the Food Standards Agency. https://www.cefas.co.uk/media/bw1f0xkg/review-of-sanitary-surveys-in-ew-2007-2011-final.pdf.

Landrigan, P.J., Stegeman, J.J., Fleming, L.E., Allemand, D., Anderson, D.M., Backer, L.C. et al. 2020. Human health and ocean pollution. Annals of Global Health 86(1): 151.

Lee, R., Kay, D., Wyer, M., Murray, L. and Stapleton, C. 2010. Sanitary profiling of shellfish harvesting areas. *In*: Rees, G., Pond, K., Kay, D., Bartram, J. and Santo Domingo, J. (eds.). Safe Management of Shellfish and Harvest Waters. IWA Publishing, London.

Lee, R.J. and Younger, A.D. 2002. Developing microbiological risk assessment for shellfish purification. International Biodeterioration and Biodegradation 50(3): 177–183.

Lees, D. 2000. Viruses and bivalve shellfish. International Journal of Food Microbiology 59: 81–116.

Leight, A.K. and Hood, R.R. 2018. Precipitation thresholds for fecal bacterial indicators in the Chesapeake Bay. Water Research 139: 252–262.

Lim, F.Y., Ong, S.L. and Hu, J. 2017. Recent advances in the use of chemical markers for tracing wastewater contamination in aquatic environment: a review. Water 9: 143.

Lozano-León, A., Rodríguez-Souto, R.R., González-Escalona, N., Taboada, J.L., Iglesias-Canle, J., Álvarez-Castro, A. et al. 2021. Detection, molecular characterization, and antimicrobial susceptibility, of *Campylobacter* spp. isolated from shellfish. Microbial Risk Analysis 18: 100176.

Lowther, J.A., Henshilwood, K. and Lees, D.N. 2008. Determination of norovirus contamination in oysters from two commercial harvesting areas over an extended period, using semiquantitative real-time reverse transcription PCR. Journal of Food Protection 71(7): 1427–1433.

Lowther, J.A., Gustar, N.E., Powell, A.L., Hartnell, R.E. and Lees, D. 2012. Two-year systematic study to assess norovirus contamination in oysters from commercial harvesting areas in the United Kingdom. Applied and Environmental Microbiology 78(16): 5812–5817.

Lowther, J.A., Cross, L., Stapleton, T., Gustar, N.E., Walker, D.I., Sills, M. et al. 2019. Use of F-specific RNA bacteriophage to estimate infectious norovirus levels in oysters. Food and Environmental Virology 11: 247–258.

Mallin, M.A., Ensign, S.H., McIver, M.R., Shank, G.C. and Fowler, P.K. 2001. Demographic, landscape, and meteorological factors controlling the microbial pollution of coastal waters. *In*: Porter, J.W. (ed.). The Ecology and Etiology of Newly Emerging Marine Diseases, Hydrobiologia, 460: 185–193. Kluwer Academic: The Netherlands.

McLeod, C., Polo, D., Le Saux, J.-C. and Le Guyader, F.S. 2017. Depuration and relaying: a review of potential removal of norovirus from oysters. Comprehensive Reviews in Food Science and Food Safety 16: 692–706.

Meghnath, K., Hasselback, P., McCormick, R., Prystajecky, N., Taylor, M., McIntyre, L. et al. 2019. Outbreaks of norovirus and acute gastroenteritis associated with British Columbia oysters, 2016–2017. Food and Environmental Virology 11: 138–148.

Ministry of Agriculture, Food Processing and Forestry. 2012. Contamination of shellfish production zones with norovirus—management framework protocol. Unpublished report.

Moon, A., Hwang, I.-G. and Choi, W.S. 2011. Prevalence of noroviruses in oysters in Korea. Food Science and Biotechnology 20(4): 1151–1154.

MPI. 2018. Animal products notice: regulated control scheme—bivalve molluscan shellfish for human consumption. Ministry for Primary Industries.

Nordgren, J., Matussek, A., Mattsson, A., Svensson, L. and Linggren, P.-E. 2009. Prevalence of norovirus and factors influencing virus concentrations during one year in a full-scale wastewater treatment plant. Water Research 43: 1117–1125.

Oliver, D.M., Porter, K.D.H., Pachepsky, Y.A., Muirhead, R.W., Reaney, S.M., Coffey, R. et al. 2016. Predicting microbial water quality with models: over-arching questions for managing risk in agricultural catchments. Science of the Total Environment 544: 39–47.

Ottoson, J., Hansen, A., Westrell, T., Johansen, K., Norder, H. and Stenström, T.A. 2006. Removal of noro- and enteroviruses, *Giardia* cysts, *Cryptosporidium* oocysts, and fecal indicators at four secondary wastewater treatment plants in Sweden. Water Environment Research 78(8): 828–834.

Pang, L., Robson, B., Farkas, K., McGill, E., Varsani, A., Gillot, L. et al. 2017. Tracking effluent discharges in undisturbed stony soil and alluvial gravel aquifer using synthetic DNA tracers. Science of the Total Environment 592: 144–152.

Payment, P., Plante, R. and Cejka, P. 2001. Removal of indicator bacteria, human enteric viruses, *Giardia* cysts, and *Cryptosporidium* oocysts at a large wastewater primary treatment plant. Canadian Journal of Microbiology 47(3): 188–193.

Pintó, R.M., Costafreda, M.I. and Bosch, A. 2009. Risk assessment in shellfish-borne outbreaks of Hepatitis A. Applied and Environmental Microbiology 75(23): 7350–7355.

Pouillot, R., Van Doren, J.M., Woods, J., Plante, D., Smith, M., Goblick, G.N. et al. 2015. Meta-analysis of the reduction of norovirus and male-specific coliphage concentrations in wastewater treatment plants. Applied Environmental Microbiology 81(14): 4669–4681.

Pouillot, R., Smith, M., van Doren, J.M., Catford, A., Holtzman, J., Calci, K.R. et al. 2022. Risk assessment of norovirus illness from consumption of raw oysters in the United States and in Canada. Risk Analysis 42(2): 344–369.

Rajko-Nenow, P., Waters, A., Keaveney, S., Flannery, J., Tuite, G., Coughlan, S. et al. 2013. Norovirus genotypes present in oysters and in effluent from a wastewater treatment plant during the seasonal peak of infections in Ireland. Applied and Environmental Microbiology 79: 2578–2587.

Reid, R.A. and Durance, T.D. 2001. The U.S. National Shellfish Sanitation Program. pp. 321–338. *In*: Hui, Y.H., Kitts, D. and Stanfield, P.S. (eds.). Foodborne Disease Handbook. Volume IV: Seafood and Environmental Toxins. CRC Press, Boca Raton, USA.

Richards, G.P. 1985. Outbreaks of shellfish-associated enteric virus illness in the United States: requisite for development of viral guidelines. Journal of Food Protection 48(9): 815–823.

Richards, G.P. 2016. Shellfish-associated enteric virus illness: virus localization, disease outbreaks and prevention. Chapter 7. pp. 185–207. *In*: Goyal, S.M. and Cannon, J.L. (eds.). Viruses in Foods. Food Microbiology and Food Safety. Springer, Switzerland.

Rincé, A., Balière, C., Hervio-Heath, D., Cozien, J., Lozach, S., Parnadeau, S. et al. 2018. Occurrence of bacterial pathogens and human noroviruses in shellfish-harvesting areas and their catchments in France. Frontiers in Microbiology 9: 2443.

Rippey, S.R. 1994. Infectious diseases associated with molluscan shellfish consumption. Clinical Microbiology Reviews 7: 419–425.

Rochelle-Newall, E., Nguyen, T.M.H., Le, T.P.Q., Sengtaheuanghoung, O. and Ribolzi, O. 2015. A short review of fecal indicator bacteria in tropical aquatic ecosystems: knowledge gaps and future directions. Frontiers in Microbiology 6: 308.

Rockey, N., Young, S., Kohn, T., Pecson, B., Wobus, C.E., Raskin, L. and Wigginton, K.R. 2020. UV disinfection of human norovirus: evaluating infectivity using a genome-wide PCR-based approach. Environmental Science and Technology 54(5): 2851–2858.

Rodríguez, R.A., Gundy, P.M., Rijal, G.K. and Gerba, C.P. 2012. The impact of combined sewer overflows on the viral contamination of receiving waters. Food and Environmental Virology 4: 34–40.

Romalde, J.L., Area, E., Sánchez, G., Ribao, C., Torrado, I., Abad, X. et al. 2002. Prevalence of enterovirus and hepatitis A virus in bivalve molluscs from Galicia (NW Spain): inadequacy of the EU standards of microbiological quality. International Journal of Food Microbiology 74: 119–130.

Rupnik, A., Keaveney, S., Devilly, L., Butler, F. and Doré, W. 2018. The impact of winter relocation and depuration on norovirus concentrations in Pacific oysters harvested from a commercial production site. Food and Environmental Virology 10: 288–296.

Schmidt, W., Evers-King, H.L., Campos, C.J.A., Jones, D.B., Miller, P.I., Davidson, K. et al. 2018. A generic approach for the development of short-term predictions of *Escherichia coli* and biotoxins in shellfish. Aquaculture Environment Interactions 10: 173–185.

Schmidt, W., Raymond, D., Parish, D., Ashton, I.G.C., Miller, P.I., Campos, C.J.A. et al. 2018a. Design and operation of a low-cost and compact autonomous buoy system for use in coastal aquaculture and water quality monitoring. Aquaculture Engineering 80: 28–36.

Scott, T.M., Rose, J.B., Jenkins, T.M., Farrah, S.R. and Lukasic, J. 2002. Microbial source tracking: current methodology and future directions. Applied and Environmental Microbiology 68(12): 5796–5803.

Sima, L.C., Schaeffer, J., Le Saux, J.-C., Parnaudeau, S., Elimelech, M. and Le Guyader, F.S. 2011. Calicivirus removal in a membrane bioreactor wastewater treatment plant. Applied and Environmental Microbiology 77: 5170–5177.

Simhon, A., Pileggi, V., Flemming, C.A., Bicudo, J.R., Lai, G. and Manoharan, M. 2019. Enteric viruses in municipal wastewater effluent before and after disinfection with chlorine and ultraviolet light. Journal of Water and Health 17(5): 670–682.

Simmons, F.J., Kuo, D.H. and Xagoraraki, I. 2011. Removal of human enteric viruses by a full-scale membrane bioreactor during municipal wastewater processing. Water Research 45: 2739–2750.

Simpson, J.M., Santo Domingo, J.W. and Resoner, D.J. 2002. Microbial source tracking: state of the science. Environmental Science & Technology 36(24): 5279–5288.

Smith, A.J., McCarthy, N., Saldana, L., Ihekweazu, C., McPhedran, K., Adak, G.K. et al. 2012. A large foodborne outbreak of norovirus in diners at a restaurant in England between January and February 2009. Epidemiology and Infection 140(9): 1695–1701.

Stapleton, C.M., Wyer, M.D., Crowther, J., McDonald, A.T., Kay, D., Greaves, J. et al. 2008. Quantitative catchment profiling to apportion faecal indicator organism budgets for the Ribble system, the UK's sentinel drainage basin for Water Framework Directive research. Journal of Environmental Management 87: 535–550.

Stapleton, C.M., Kay, D., Magill, S.H., Wyer, M.D., Davies, C., Watkins, J. et al. 2011. Quantitative microbial source apportionment as a tool in aiding the identification of microbial risk factors in shellfish harvesting waters: the Loch Etive case study. Aquaculture Research 42: 1–20.

Strubbia, S., Schaeffer, J., Munnink, B.D.O., Besnard, A., Phan, M.V.T., Nieuwenhuijse, D.F. et al. 2019. Metavirome sequencing to evaluate norovirus diversity in sewage and related bioaccumulated oysters. Frontiers in Microbiology 10: 2394.

Tetra Tech and Herrera Environmental Consultants. 2011. Using microbial source tracking to support TMDL development and implementation. Report prepared for the USEPA. https://www.epa.gov/tmdl/using-microbial-source-tracking-support-tmdl-development-and-implementation.

Torok, V., Hodgson, K., McLeod, C., Tan, J. and Malhi, N. 2018. National survey of foodborne viruses in Australian oysters at production. Food Microbiology 69: 196–203.

True, E.D. 2018. Using a numerical model to track the discharge of a wastewater treatment plant in a tidal estuary. Water Air and Soil Pollution 229: 267.

USEPA. 2015. Review of coliphages as possible indicators of fecal contamination for ambient water quality. Report 820-R-15-098 of the EPA Office of Water. https://www.epa.gov/sites/default/

files/2016-07/documents/review_of_coliphages_as_possible_indicators_of_fecal_contamination_for_ambient_water_quality.pdf.

USFDA and ISSC. 2019. National shellfish sanitation program (NSSP): guide for the control of molluscan shellfish, 2019 Revision. https://www.fda.gov/media/143238/download.

van Dolah, R.F., Riekerk, G.H.M., Bergquist, D.C., Felber, J., Chestnut, D.E. and Holland, A.F. 2008. Estuarine habitat quality reflects urbanization at large spatial scales in South Carolina's coastal zone. Science of the Total Environment 390: 142–154.

Verhougstraete, M.P., Martin, S.L., Kendall, A.D., Hyndman, D.W. and Rose, J.B. 2015. Linking fecal bacteria in rivers to landscape, geochemical, and hydrologic factors and sources at the basin scale. Proceedings of the National Academy of Sciences 112(33): 10419–10424.

Walker, D.I., Fok, B.C.T. and Ford, C.L. 2020. A qPCR-MPN method for rapid quantification of *Escherichia coli* in bivalve molluscan shellfish. Journal of Microbiological Methods 178: 106067.

Wan, V., McIntyre, L., Kent, D., Leong, D. and Henderson, S.B. 2018. Near-real-time surveillance of illnesses related to shellfish consumption in British Columbia: analysis of Poison Center data. JMIR Public Health Surveillance 4(1): e17.

Wang, D., Wu, Q., Kou, X., Yao, L. and Zhang, J. 2008. Distribution of norovirus in oyster tissues. Journal of Applied Microbiology 105: 1966–1972.

Wang, J. and Deng, Z. 2019. Modeling and predicting fecal coliform bacteria levels in oyster harvest waters along Louisiana Gulf coast. Ecological Indicators 101: 212–220.

Webber, J.L., Tyler, C.R., Carless, D., Jackson, B., Tingley, D., Stewart-Sinclair, P. et al., 2021. Impacts of land use on water quality and the viability of bivalve shellfish mariculture in the UK: a case study and review for SW England. Environmental Science & Policy 126: 122–131.

WHO. 2016. Quantitative microbial risk assessment: application for water safety management. World Health Organization, Geneva.

Winterbourn, J.B., Clements, K., Lowther, J.A., Malham, S.K., McDonald, J.E. and Jones, D.L. 2016. Use of *Mytilus edulis* biosentinels to investigate spatial patterns of norovirus and faecal indicator organism contamination around coastal sewage discharges. Water Research 105: 241–250.

Wither, A., Greaves, J., Dunhill, I., Wyer, M., Stapleton, C., Kay, D. et al. 2005. Estimation of diffuse and point source microbial pollution in the Ribble catchment discharging to bathing waters in the north west of England. Water Science and Technology 51(3–4): 191–198.

Wyer, M.D., Kay, D., Watkins, J., Davies, C., Kay, C., Thomas, R. et al. 2010. Evaluating short-term changes in recreational water quality during a hydrograph event using a combination of microbial tracers, environmental microbiology, microbial source tracking and hydrological techniques: a case study in southwest Wales, UK. Water Research 44: 4783–4795.

Wyer, M.D., Kay, D., Morgan, H., Naylor, S., Clark, S., Watkins, J. et al. 2018. Within-day variability in microbial concentrations at a UK designated bathing water: implications for regulatory monitoring and the application of predictive modelling based on historical compliance data. Water Research X 1: 100006.

Younger, A.D. and Reese, R.A. 2013. Comparison of *Escherichia coli* levels between bivalve mollusc species across harvesting sites in England and Wales. Journal of Shellfish Research 32(2): 527–532.

Zhang, Q., Gallard, J., Wu, B., Harwood, V.J., Sadowsky, M.J., Hamilton, K.A. et al. 2019. Synergy between quantitative microbial source tracking (qMST) and quantitative microbial risk assessment (QMRA): a review and prospectus. Environment International 130: 104703.

Zimmer-Faust, A.G., Brown, C.A. and Manderson, A. 2018. Statistical models of fecal coliform levels in Pacific northwest estuaries for improved shellfish harvest area closure decision making. Marine Pollution Bulletin 137: 360–369.

Chapter 8

Virus Removal and Inactivation

Monika Trzaskowska,[1] *Kevin Hunt*[2] and *David Rodríguez-Lázaro*[3,4,]*

1. Introduction

Viruses are now recognized as a major foodborne hazard, with enteric viruses causing a large fraction of the foodborne disease burden (FAO and WHO 2008, Hall et al. 2014, Havelaar et al. 2015). Unlike bacteria, viruses will not grow on food once contaminated, and pose little risk of spoilage. Because the most common viruses can be infectious at low doses, it is important to ensure that they are removed or inactivated before consumption (Bradshaw and Jaykus 2016).

Removal and inactivation are control strategies used to manage food-related risks. They should be distinguished from survival and persistence of the viruses in foods. The high rate of persistence among foodborne viruses makes inactivation methods even more critical to understand (Sánchez and Bosch 2016). Preservation methods intended to slow bacterial growth can have the opposite effect on viruses, allowing them to persist even longer. Higher impact methods are needed to manage virus risk. These can broadly be divided between physical and chemical processes. Physical processes themselves can be divided between thermal treatments and nonthermal, such as high hydrostatic pressure or irradiation. The efficacy of these techniques on reducing virus load, as shown in the literature, is the focus of this chapter. Until recently, laboratory cultivation of enteric viruses was not possible, and data from surrogate viruses was necessary for any estimation of inactivation. This is

---
[1] Institute of Human Nutrition Sciences, Department of Food Gastronomy and Food Hygiene, Warsaw University of Life Sciences–SGGW, Nowoursynowska 159c Str., 02-776 Warsaw, Poland.
[2] Centre for Food Safety, School of Biosystems and Food Engineering, University College Dublin, Belfield, Dublin 4, Ireland.
[3] Microbiology Division. Department of Biotechnology and Food Science, Faculty of Sciences, University of Burgos, Plaza Misael Bañuelos s/n, 09001 Burgos, Spain.
[4] Research Centre for Emerging pathogens and Global Health. University of Burgos, Burgos, Spain.
* Corresponding author: drlazaro@ubu.es

an unavoidable limitation on the data, as issues with suitability and representativeness of surrogates are well known (Baert et al. 2009, Knight et al. 2016). The greater availability of more accurate detection methods for human noroviruses can provide more accurate data (Ettayebi et al. 2021).

1.1 Physical processes

1.1.1 Thermal processes

Table 1 shows the effect of thermal processing on foodborne viruses and their surrogates. Data on foodborne matrices has been prioritized.

For rotavirus, heating at temperatures of 60–72°C for a period of 10–30 minutes showed over 6 log reductions in cell culture media (Araud et al. 2016, O' Mahony et al. 2000). However, inactivation in cell culture media can be more effective than in a food matrix. A HAV suspension heated for 60°C for 10 minutes and 80°C for 3 minutes showed log reductions of > 4.6. Similar temperature and time conditions had only a 1 and 2-log reductions in strawberries and shellfish homogenates, respectively (Croci et al. 1999, Deboosere et al. 2004). The rate of reduction was similar at temperatures below 80°C, with temperatures over 90°C or needed to increase over a 3 log reduction (Butot et al. 2009, Harlow et al. 2011, Sow et al. 2011).

HEV is an emerging pathogen in food, for which very little inactivation data is available. To reduce the risk of HEV, food should be cooked thoroughly. Temperatures of 60–70°C for 5–15 minutes results in a reduction of 1–3 log in liver and pork meat (Barnaud et al. 2012, Imagawa et al. 2018).

Data for human noroviruses is sparse, and indicates high resistance even to boiling temperature, with only a 2 log reduction after 3 minutes (Hewitt and Greening 2006). However, using a human norovirus surrogate, feline calicivirus, higher reductions were observed. 6 log reduction after 3–5 minutes at 70°C in cell culture media (Doultree et al. 1999), or less than 3 log reduction in raspberry and clam with temperatures of 65–75°C (Baert et al. 2008, Sow et al. 2011). However the inability to distinguish between infectious and non-infectious viral particles can underestimate the real virus inactivation. Better detection methods show a 2.5 log reduction after 2 minutes at 90°C, a higher rate for a lower temperature and time profile (Li et al. 2017).

1.2 Nonthermal processes

1.2.1 High hydrostatic pressure treatment (HPP)

High hydrostatic pressure (HPP) can be applied to foods where the effect of thermal processing has a severe impact on food sensorial properties (Murchie et al. 2007). Food is submerged in liquid, typically water, at high pressures, for controlled temperature and times, and the virus capsid is disrupted with minimal impact on the food matrix itself. The effect is instant and homogenous across the matrix. At higher pressures, changes in food shape and structure can be observed (Hoover et al. 1989). The significant factors in optimizing the process are pressure, temperature and time under pressure. Since its development, HPP has been successfully applied

Table 1. The effect of thermal treatment on various viruses and matrices.

Family	Virus	Matrix	Temp. (°C)	Time (min)	Log_{10} reduction	Reference
Reoviridae	Human rotavirus	Media	60	10	7.0	(O' Mahony et al. 2000)
Reoviridae	Human rotavirus	Media	62	30	> 2.0	(Araud et al. 2016)
Reoviridae	Human rotavirus	Media	72	1	> 6.0	(Araud et al. 2016)
Picornaviridae	Hepatitis A virus	Milk	85	< 0.5	5.0	(Bidawid et al. 2000)
Picornaviridae	Hepatitis A virus	Milk (skimmed)	71	6.55	4.0	(Bidawid et al. 2000)
Picornaviridae	Hepatitis A virus	Milk (homogenized)	71	8.31	4.0	(Bidawid et al. 2000)
Picornaviridae	Hepatitis A virus	Milk	63	1	1.6	(Hewitt et al. 2009)
Picornaviridae	Hepatitis A virus	Milk	72	1	2.2	(Hewitt et al. 2009)
Picornaviridae	Hepatitis A virus	Milk	62.8	30	3.0	(Parry and Mortimer 1984)
Picornaviridae	Hepatitis A virus	Milk	71.6	0.25	2.0	(Parry and Mortimer 1984)
Picornaviridae	Hepatitis A virus	Cream	71	12.67	4.0	(Bidawid et al. 2000)
Picornaviridae	Hepatitis A virus	Basil	95	2.5	> 3.0	(Butot et al. 2009)
Picornaviridae	Hepatitis A virus	Basil	75	2.5	1.87	(Butot et al. 2009)
Picornaviridae	Hepatitis A virus	PBS	60	74.6	1.0	(Gibson and Schwab 2011)
Picornaviridae	Hepatitis A virus	Clam	90	3	4.75	(Sow et al. 2011)
Picornaviridae	Hepatitis A virus	Mussel	Steam	6	> 3.0	(Harlow et al. 2011)
Picornaviridae	Hepatitis A virus	Shellfish homogenate	60	10	2.0	(Croci et al. 1999)
Picornaviridae	Hepatitis A virus	Shellfish homogenate	80	3	2	(Croci et al. 1999)
Picornaviridae	Hepatitis A virus	Clam homogenate	60	5	1.0	(Bozkurt et al. 2015a)
Picornaviridae	Hepatitis A virus	PBS	60	10	> 3.0	(Bozkurt et al. 2014a)
Picornaviridae	Hepatitis A virus	Viral suspension	60	10	> 4.6	(Croci et al. 1999)
Picornaviridae	Hepatitis A virus	Viral suspension	80	3	> 4.6	(Croci et al. 1999)

Table 1 contd. ...

...Table 1 contd.

Family	Virus	Matrix	Temp. (°C)	Time (min)	Log_{10} reduction	Reference
Picornaviridae	Hepatitis A virus	Strawberry homogenate	85	0.96	1.0	(Deboosere et al. 2004)
Picornaviridae	Hepatitis A virus	Strawberry homogenate	85	4.98	1.0	(Deboosere et al. 2004)
Picornaviridae	Hepatitis A virus	Strawberry homogenate	80	8.94	1.0	(Deboosere et al. 2004)
Picornaviridae	Hepatitis A virus	Green onions	65.9	20 hrs	3.9	(Laird et al. 2011)
Picornaviridae	Hepatitis A virus	Spinach	65	6	> 2.0	(Bozkurt et al. 2015c)
Picornaviridae	Hepatitis A virus	Turkey deli meat	65	1.5	< 1.0	(Bozkurt et al. 2015b)
Picornaviridae	Hepatitis A virus	Turkey deli meat	72	1	1.0	(Bozkurt et al. 2015b)
Picornaviridae	Poliovirus	Milk	72	0.25	0.56	(Strazynski et al. 2002)
Picornaviridae	Poliovirus	Milk	72	0.5	> 5.0	(Strazynski et al. 2002)
Picornaviridae	Poliovirus	Oysters	Steam	30	2.0	(DiGirolamo et al. 1970)
Adenoviridae	Adenovirus type 5	Media	85	2 hrs	> 5.5	(Sauerbrei and Wutzler 2009)
Hepeviridae	Hepatitis E virus	Stool	60	1 hr	> 5.0	(Emerson et al. 2005)
Hepeviridae	Hepatitis E virus	Liver homogenate	62	5	1.2	(Barnaud et al. 2012)
Hepeviridae	Hepatitis E virus	Liver homogenate	71	10	2.6	(Barnaud et al. 2012)
Hepeviridae	Hepatitis E virus	Pork	70	5	> 3.0	(Imagawa et al. 2018)
Caliciviridae	Human norovirus	Stool	60	15	> 5.0	(Ettayebi et al. 2016)
Caliciviridae	Feline calicivirus	Media	71.3	1	3.0	(Duizer et al. 2004)
Caliciviridae	Feline calicivirus	Media	37	24 hrs	3.0	(Duizer et al. 2004)
Caliciviridae	Feline calicivirus	Media	56	8	3.0	(Duizer et al. 2004)
Caliciviridae	Feline calicivirus	PBS	60	14.1	1.0	(Gibson and Schwab 2011)
Caliciviridae	Feline calicivirus	Petfood	79	0.5	> 4.4	(Haines et al. 2015)
Caliciviridae	Feline calicivirus	Basil	95	2.5	> 4.0	(Butot et al. 2009)
Caliciviridae	Feline calicivirus	Basil	75	2.5	3.98	(Butot et al. 2009)

Table 1 contd. ...

...Table 1 contd.

Family	Virus	Matrix	Temp. (°C)	Time (min)	Log_{10} reduction	Reference
Caliciviridae	Feline calicivirus	Cockles	Boil	0.5	1.7	(Slomka and Appleton 1998)
Caliciviridae	Feline calicivirus	Media	56	3	None	(Doultree et al. 1999)
Caliciviridae	Feline calicivirus	Media	56	60	7.5	(Doultree et al. 1999)
Caliciviridae	Feline calicivirus	Media	70	1	3.0	(Doultree et al. 1999)
Caliciviridae	Feline calicivirus	Media	70	3	6.5	(Doultree et al. 1999)
Caliciviridae	Feline calicivirus	Media	70	5	7.5	(Doultree et al. 1999)
Caliciviridae	Feline calicivirus	Media	Boil	1	7.5	(Doultree et al. 1999)
Caliciviridae	Feline calicivirus	Media	70	1.5	6.0	(Buckow et al. 2008)
Caliciviridae	Feline calicivirus	Media	63	0.41	1.0	(Cannon et al. 2006)
Caliciviridae	Feline calicivirus	Media	72	0.12	1.0	(Cannon et al. 2006)
Caliciviridae	Murine norovirus 1	Media	80	2.5	6.5	(Baert et al. 2008)
Caliciviridae	Murine norovirus 1	PBS	60	13.7	1.0	(Gibson and Schwab 2011)
Caliciviridae	Murine norovirus 1	Clam	90	1.5	5.47	(Sow et al. 2011)
Caliciviridae	Murine norovirus 1	Raspberry homogenate	65	0.5	1.86	(Baert et al. 2008)
Caliciviridae	Murine norovirus 1	Raspberry homogenate	75	0.25	2.81	(Baert et al. 2008)
Caliciviridae	Murine norovirus 1	Media	63	0.44	1.0	(Cannon et al. 2006)
Caliciviridae	Murine norovirus 1	Media	72	0.17	1.0	(Cannon et al. 2006)
Caliciviridae	Feline calicivirus	Media	60	2	> 4.5	(Topping et al. 2009)
Caliciviridae	Feline calicivirus	Media	65	2	> 5.0	(Topping et al. 2009)
Caliciviridae	Feline calicivirus	PBS	60	5	> 5.0	(Bozkurt et al. 2014a)
Caliciviridae	Feline calicivirus	Mussel homogenate	60	1	4.9	(Bozkurt et al. 2014b)
Caliciviridae	Feline calicivirus	Mussel homogenate	65	0.5	> 7	(Bozkurt et al. 2014b)

Table 1 contd. ...

...Table 1 contd.

Family	Virus	Matrix	Temp. (°C)	Time (min)	Log$_{10}$ reduction	Reference
Caliciviridae	Feline calicivirus	Turkey deli meat	65	0.5	> 6.0	(Bozkurt et al. 2015b)
Caliciviridae	Feline calicivirus	Turkey deli meat	72	0.5	> 6.0	(Bozkurt et al. 2015b)
Caliciviridae	Human norovirus	Mussel	Boil	3	2.0	(Hewitt and Greening 2006)
Caliciviridae	Human norovirus	Media	90	2	2.5	(Li et al. 2017)
Caliciviridae	Murine norovirus	PBS	60	5	> 5.0	(Bozkurt et al. 2014a)
Caliciviridae	Murine norovirus	Mussel homogenate	60	1	2.2	(Bozkurt et al. 2014b)
Caliciviridae	Murine norovirus	Mussel homogenate	72	0.33	> 6.0	(Bozkurt et al. 2014b)
Caliciviridae	Murine norovirus	Turkey deli meat	65	0.5	> 5.0	(Bozkurt et al. 2015b)
Caliciviridae	Murine norovirus	Turkey deli meat	72	0.5	> 5.0	(Bozkurt et al. 2015b)
Caliciviridae	Murine norovirus	Oyster homogenate	63	2	> 3.0	(Shao et al. 2018)
Caliciviridae	Murine norovirus	Oyster homogenate	67	1	> 5.0	(Shao et al. 2018)
Caliciviridae	Tulane virus	Oyster homogenate	63	0.5	> 2.0	(Shao et al. 2018)
Caliciviridae	Tulane virus	Oyster homogenate	63	1	> 3.0	(Shao et al. 2018)
Caliciviridae	Tulane virus	Media	56	10	> 6.0	(Ailavadi et al. 2019)

commercially on fruit and vegetable products and ready to eat seafood (Lee and Grove 2016).

Experiments in hepatitis A virus (HAV) in cell culture media (Table 2) show a high degree of variance, with reductions ranging from < 0.5 to > 6 (400–450 MPa, ambient temperatures, 5–10 min) (Grove et al. 2008, Kingsley et al. 2002). Processing at low temperatures shows high reduction rates of over 4.6 log reduction in oysters (Kingsley and Chen 2009). Similar results are shown for reducing HAV in strawberry homogenate (4.3 log) and green onions (5.5 external, 2.5 internal) (Hirneisen and Kniel 2013, Kingsley et al. 2005)

For the calicivirus family (human noroviruses and its surrogates), the reductions observed are low, with values of 1 log or less for norovirus in strawberries (DiCaprio et al. 2019). Higher reductions are seen for FCV and MNV in Table 2, but still appearing less than for HAV. It is challenging to compare effects across different hazards, food matrices and pressure and temperature profiles, and some degree of optimization is essential in applying this method.

Table 2. The effect of high hydrostatic pressure (HPP) treatment on various viruses and matrices.

Family	Virus	Treatment	Matrix	Log_{10} reduction	Reference
Reoviridae	Rotavirus	300 MPa, 25°C, 2 min	Media	8	(Khadre and Yousef 2002)
Picornaviridae	Hepatitis A virus	450 MPa, ambient temp, 5 min	Media	> 6	(Kingsley et al. 2002)
Picornaviridae	Hepatitis A virus	400 MPa, ambient temp, 10 min	Media	2	(Grove et al. 2008)
Picornaviridae	Hepatitis A virus	400 MPa, ambient temp, 10 min	Media	< 0.5	(Grove et al. 2008)
Picornaviridae	Hepatitis A virus	400 MPa, 9°C, 1 min	Oysters	3	(Calci et al. 2005)
Picornaviridae	Hepatitis A virus	375 MPa, 21°C, 5 min	Strawberry homogenate	4.3	(Kingsley et al. 2005)
Picornaviridae	Hepatitis A virus	375 MPa, 21°C, 5 min	Green onions	4.7	(Kingsley et al. 2005)
Picornaviridae	Hepatitis A virus	500 MPa, 4°C, 5 min	Sausage	3.23	(Sharma et al. 2008)
Picornaviridae	Hepatitis A virus	500 MPa, 2°C, 1 min, pH 6.07	Oyster	> 4.6	(Kingsley and Chen 2009)
Picornaviridae	Hepatitis A virus	500 MPa, 4°C, 1 min, pH 6.07	Oyster	> 4.6	(Kingsley and Chen 2009)
Picornaviridae	Hepatitis A virus	500 MPa, 5°C, 1 min, pH 6.07	Oyster	4.6	(Kingsley and Chen 2009)
Picornaviridae	Hepatitis A virus	500 MPa, 2°C, 2 min	Green onions (internal)	2.5	(Hirneisen and Kniel 2013, p. 201)
Picornaviridae	Hepatitis A virus	500 MPa, 2°C, 2 min	Green onions (external)	5.5	(Hirneisen and Kniel 2013, p. 201)
Picornaviridae	Poliovirus	600 MPa, ambient temp, 5 min	Media	None	(Kingsley et al. 2004)
Picornaviridae	Poliovirus	600 MPa, ambient temp, 5 min	Media	None	(Wilkinson et al. 2001)
Picornaviridae	Poliovirus	600 MPa, ambient temp, 5 min	Media	None	(Grove et al. 2008)
Picornaviridae	Aichivirus	600 MPa, ambient temp, 5 min	Media	None	(Kingsley et al. 2004)
Picornaviridae	Coxsackievirus B5	600 MPa, ambient temp, 5 min	Media	None	(Kingsley et al. 2004)
Picornaviridae	Coxsackievirus A	600 MPa, ambient temp, 5 min	Media	7.6	(Kingsley et al. 2004)
Caliciviridae	Feline calicivirus	275 MPa, ambient temp, 5 min	Media	> 6	(Kingsley et al. 2002)
Caliciviridae	Feline calicivirus	200 MPa, − 10°C, 4 min	Media	5.0	(Chen et al. 2005)
Caliciviridae	Feline calicivirus	200 MPa, 20°C, 4 min	Media	0.3	(Chen et al. 2005)
Caliciviridae	Feline calicivirus	300 MPa, ambient temp, 3 min	Media	5	(Grove et al. 2008)

Table 2 contd. ...

...Table 2 contd.

Family	Virus	Treatment	Matrix	Log$_{10}$ reduction	Reference
Caliciviridae	Feline calicivirus	500 MPa, 4°C, 5 min	Sausage	2.89	(Sharma et al. 2008)
Caliciviridae	Feline calicivirus	250 MPa, 20 C, 5 min	Oyster	1.5	(Murchie et al. 2007)
Caliciviridae	Feline calicivirus	250 MPa, 20 C, 5 min	Mussel	1.5	(Murchie et al. 2007)
Caliciviridae	Feline calicivirus	300 MPa, 20 C, 5 min	Oyster	> 3.83	(Murchie et al. 2007)
Caliciviridae	Feline calicivirus	400 MPa, ambient temp, 10 min	Pig liver	> 5.0	(Emmoth et al. 2017)
Caliciviridae	Feline calicivirus	600 MPa, ambient temp, 10 min	Ham	4.4	(Emmoth et al. 2017)
Caliciviridae	Murine norovirus 1	400 MPa, 5°C, 5 min	Oyster	4	(Kingsley et al. 2007)
Caliciviridae	Murine norovirus 1	450 MPa, 20°C, 5 min	Media	6.85	(Kingsley et al. 2007)
Caliciviridae	Murine norovirus 1	350 MPa, 4°C, 2 min	Media	8.1	(Lou et al. 2011)
Caliciviridae	Murine norovirus 1	350 MPa, 4°C, 2 min	Lettuce	2.4	(Lou et al. 2011)
Caliciviridae	Murine norovirus 1	350 MPa, 4°C, 2 min	Strawberry	2.2	(Lou et al. 2011)
Caliciviridae	Murine norovirus 1	400 MPa, 0°C, 5 min	Oyster	> 4	(Li et al. 2009)
Caliciviridae	Human norovirus	600 MPa, 21 C, 2 min	Blueberries	< 1.0	(Li et al. 2017, Li et al. 2013)
Caliciviridae	Human norovirus	500 MPa, 1 C, 2 min	Blueberries and water	2.7	(Li et al. 2013)
Caliciviridae	Human norovirus	400 MPa, 4 C, 2 min, pH 7.0	Strawberry puree	2.29	(DiCaprio et al. 2019)
Caliciviridae	Human norovirus	400 MPa, 20 C, 2 min, pH 7.0	Strawberry puree	1.0	(DiCaprio et al. 2019)
Caliciviridae	Human norovirus	500 MPa, 4 C, 2 min, pH 4.0	Strawberry puree	< 1.0	(DiCaprio et al. 2019)
Caliciviridae	Human norovirus	600 MPa, 4 C, 2 min, pH 4.0	Strawberry puree	None	(DiCaprio et al. 2019)
Caliciviridae	Human norovirus	400 MPa, 25 C, 5 min	Oyster	1.9	(Imamura et al. 2017)
Leviviridae	MS2	600 MPa, 21°C, 10 min	Media	1.5	(Guan et al. 2006)
Leviviridae	MS2	500 MPa, 4°C, 5 min	Sausage	1.47	(Sharma et al. 2008)
Leviviridae	MS2	600 MPa, 10 min, 13°C	Cured ham	1.3	(Emmoth et al. 2017)
Hepeviridae	Hepatitis E virus	400 MPa, ambient temp, 5 min	Pork paté	0.5	(Nasheri et al. 2020)
Hepeviridae	Hepatitis E virus	600 MPa, ambient temp, 5 min	Pork paté	0.5	(Nasheri et al. 2020)

Data on HHP inactivation for HEV is still very scarce. Hence, more data is needed to assess the potential use of HPP as an inactivation treatment for HEV in food. For example, HHP inactivation for HEV in pork paté has shown a very low reduction, as few as 0.5 for 600 MPa (Nasheri et al. 2020).

1.2.2 Irradiation

Irradiation in various forms has long been proven an effective nonthermal process for preserving foods. Viruses are known to be more resistant to radiation treatment than bacteria and other biological hazards. Higher log reductions have been shown when combined with sanitization treatments (Foley et al. 2002).

Gamma irradiation is a form of ionizing radiation, approved in doses up to 4 kGy for fresh produce by the US FDA. Viruses are more resistant to this than bacteria, due to their structure and size (Hirneisen et al. 2010) (Table 3). The effect of the food matrix, in particular the presence of proteins, has a strong inhibitory impact on the inactivation effect (de Roda Husman et al. 2004). In lettuce, strawberry and oyster, doses below 4 kGy do not achieve more than a 1 log reduction for HAV (Bidawid et al. 2000, Mallett et al. 1991), while a 7 log reduction was obtained for a suspension of MS2 with the same dose (de Roda Husman et al. 2004).

Electron beam (Table 4) is another ionizing radiation-based technology that has been explored, with similar conclusions. Among calicivirus surrogates, murine norovirus shows greater resistance than Tulane virus (Praveen et al. 2013, Predmore et al. 2015). In strawberries, MS2 was reduced less than 1 log when exposed to 4 kGy (Jung et al. 2009), which compares harshly with the effect shown in suspension (de Roda Husman et al. 2004).

Table 3. The effect of gamma irradiation treatment on various viruses and matrices.

Family	Virus	Treatment	Matrix	Log_{10} reduction	Reference
Reoviridae	Rotavirus	2.4 kGy	Oyster	1.0	(Mallett et al. 1991)
Reoviridae	Rotavirus	2.4 kGy	Clams	1.0	(Mallett et al. 1991)
Picornaviridae	Hepatitis A virus	3 kGy	Lettuce	1.0	(Bidawid et al. 2000)
Picornaviridae	Hepatitis A virus	3 kGy	Strawberry	1.0	(Bidawid et al. 2000)
Picornaviridae	Hepatitis A virus	2 kGy	Oyster	1.0	(Mallett et al. 1991)
Picornaviridae	Hepatitis A virus	2 kGy	Clam	1.0	(Mallett et al. 1991)
Picornaviridae	Poliovirus	2.84 kGy	Oyster	1	(Jung et al. 2009)
Picornaviridae	Coxsackievirus B2	7 kGy	Beef	1	(Sullivan et al. 1973)
Caliciviridae	Human norovirus	8 kGy	Stool	> 5.0	(Ettayebi et al. 2016)
Caliciviridae	Feline calicivirus	200 Gy	Suspension	1.6	(de Roda Husman et al. 2004)
Caliciviridae	Murine norovirus	2.8 kGy	Strawberry	1.3	(Feng et al. 2011)
Caliciviridae	Murine norovirus	5.6 kGy	Strawberry	2.4	(Feng et al. 2011)
Leviviridae	MS2	2 kGy	Suspension	7.0	(de Roda Husman et al. 2004)

Table 4. The effect of electron beam irradiation treatment on various viruses and matrices.

Family	Virus	Treatment	Matrix	Log$_{10}$ reduction	Reference
Caliciviridae	Murine norovirus	2 kGy (e-beam)	PBS	< 1.0	(Praveen et al. 2013)
Caliciviridae	Murine norovirus	4–12 kGy (e-beam)	PBS	< 6.4	(Praveen et al. 2013)
Caliciviridae	Murine norovirus	4 kGy (e-beam)	Cabbage	1.0	(Praveen et al. 2013)
Caliciviridae	Murine norovirus	12 kGy (e-beam)	Cabbage	< 3.0	(Praveen et al. 2013)
Caliciviridae	Murine norovirus	6 kGy (e-beam)	Strawberry	< 1.0	(Praveen et al. 2013)
Caliciviridae	Murine norovirus	12 kGy (e-beam)	Strawberry	2.2	(Praveen et al. 2013)
Caliciviridae	Tulane virus	16 kGy (e-beam)	Strawberry	7.0	(Predmore et al. 2015)
Caliciviridae	Tulane virus	16 kGy (e-beam)	Lettuce	7.0	(Predmore et al. 2015)
Leviviridae	MS2	4 kGy (e-beam)	Strawberry	< 1	(Jung et al. 2009)

1.2.3 UV irradiation and pulsed light

UV (Table 5) is a non-ionizing source of radiation technology, based on light. Its use for virus control is limited on the surfaces of food, and does not inactivate internal viruses in produce or viruses shielded by textures or cracks in the surface (Bosch et al. 2018). Pulsed light (Table 6) is similar in application, using a broad spectrum of light that includes 45% UV. It exposes produce to short pulses of high intensity, with high efficacy in liquid suspension (Roberts and Hope 2003, Vimont et al. 2014) and less efficacy in food matrices (Huang et al. 2017, Huang and Chen 2015).

1.2.4 Cold plasma

Cold plasma, or nonthermal plasma, is an emerging technology demonstrating high potential at reducing viral loads from oysters and produce (Table 7). It is rapid, does not affect the quality of the food without toxic residues. It is based on the generation of plasma in the atmosphere, caused by electrical discharges (Molteni and Donazzi 2020). There are many parameters involved, and the mechanism of effect on the virus is not fully understood; possibly to the effect on both the capsid envelope and nucleic acid phosphate backbone (Aman Mohammadi et al. 2021). A 60 minute treatment showed a reduction greater than 4.5 log in raw oysters (Choi et al. 2020), which is significantly faster and more effective than the most common method of shellfish treatment, depuration (Table 7). However, as for HPP, the process will result in the death of the oyster, limiting its implementation in the shellfish industry.

Table 5. The effect of UV light treatment on various viruses and matrices.

Family	Virus	Treatment	Matrix	Log_{10} reduction	Reference
Picornaviridae	Hepatitis A virus	40 mW s/cm²	Lettuce	4.3	(Fino and Kniel 2008)
Picornaviridae	Hepatitis A virus	40 mW s/cm²	Green onions	4.2	(Fino and Kniel 2008)
Picornaviridae	Hepatitis A virus	40 mW s/cm²	Strawberries	1.3	(Fino and Kniel 2008)
Picornaviridae	Hepatitis A virus	120 mW s/cm²	Lettuce	4.5	(Fino and Kniel 2008)
Picornaviridae	Hepatitis A virus	120 mW s/cm²	Green onions	5.3	(Fino and Kniel 2008)
Picornaviridae	Hepatitis A virus	120 mW s/cm²	Strawberries	1.8	(Fino and Kniel 2008)
Picornaviridae	Hepatitis A virus	0.212 J/cm²	Strawberries	1.0	(Lytle and Sagripanti 2005)
Picornaviridae	Hepatitis A virus	0.13 J/cm²	Strawberries	1.0	(Ortiz-Solà et al. 2021)
Picornaviridae	Hepatitis A virus	0.212 J/cm²	Raspberries	1.0	(Park et al. 2015)
Picornaviridae	Hepatitis A virus	0.240 J/cm²	Green onions	5.0	(Hirneisen and Kniel 2013)
Picornaviridae	Hepatitis A virus	3.6 J/cm²	Chicken breast	1.0	(Park et al. 2015)
Caliciviridae	Feline calicivirus	120 mW s/cm²	Suspension	3.0	(de Roda Husman et al. 2004)
Caliciviridae	Feline calicivirus	25 mW s/cm²	Media	4.0	(Park et al. 2011)
Caliciviridae	Murine norovirus	29 mW s/cm²	Media	4.0	(Park et al. 2011)
Caliciviridae	Murine norovirus	50 mW s/cm²	PBS	2.8	(Jean et al. 2011)
Caliciviridae	Murine norovirus	50 mW s/cm²	Fetal calf serum	> 5.0	(Jean et al. 2011)
Caliciviridae	Murine norovirus	1.2 J/cm²	Blueberries, wet	> 4.3	(Liu et al. 2015)
Caliciviridae	Murine norovirus	1.2 J/cm²	Blueberries, dry	2.5	(Liu et al. 2015)
Leviviridae	MS2	650 mW s/cm²	Suspension	3.0	(de Roda Husman et al. 2004)
Caliciviridae	Murine norovirus	1.331 J/cm²	Blueberry	1.0	(Lytle and Sagripanti 2005)
Caliciviridae	Murine norovirus	1.2 J/cm²	Blueberry	5.0	(Liu et al. 2015)
Caliciviridae	Murine norovirus	0.24 J/cm²	Green onions	1.0	(Hirneisen and Kniel 2013)
Caliciviridae	Murine norovirus	0.6 J/cm²	Lettuce	1.0	(Moon et al. 2021)
Leviviridae	MS2	0.019 J/cm²	Lettuce	1.0	(Xie et al. 2008)
Leviviridae	MS2	70 mW s/cm²	Cell culture	4.0	(de Roda Husman et al. 2004, Park et al. 2011)
Caliciviridae	Murine norovirus	3.6 J/cm²	Chicken breast	1.0	(Park et al. 2015)

Table 6. The effect of pulsed light treatment on various viruses and matrices.

Family	Virus	Treatment	Matrix	Log_{10} reduction	Reference
Picornaviridae	Hepatitis A virus	10 J/cm^2	PBS	4.1	(Roberts and Hope 2003)
Picornaviridae	Hepatitis A virus	10 J/cm^2	Fetal calf serum	5.7	(Roberts and Hope 2003)
Picornaviridae	Poliovirus	10 J/cm^2	PBS	3.2	(Roberts and Hope 2003)
Picornaviridae	Poliovirus	10 J/cm^2	Fetal calf serum	6.7	(Roberts and Hope 2003)
Picornaviridae	Poliovirus	0.28 J/cm^2	PBS	4.0	(Lamont et al. 2007)
Caliciviridae	Murine norovirus	12 J/cm^2	Suspension	> 3.0	(Vimont et al. 2014)
Caliciviridae	Murine norovirus	1.27 J/cm^2	Strawberry	1.8	(Huang and Chen 2015)
Caliciviridae	Murine norovirus	22.5 J/cm^2	Blueberry	3.8	(Huang et al. 2017)
Adenoviridae	Adenovirus	5.6 J/cm^2	PBS	4.0	(Lamont et al. 2007)

Table 7. The effect of cold plasma treatment on various viruses and matrices.

Family	Virus	Treatment	Matrix	Log_{10} reduction	Reference
Caliciviridae	Human norovirus	Atmospheric, 60 min	Oyster	> 4.5	(Choi et al. 2020)
Caliciviridae	Human norovirus	Nitrogen-oxygen, 12 min	Salmon	> 4.5	(Huang et al. 2021)
Caliciviridae	Human norovirus	Nitrogen-oxygen, 30 min	Clam meat	1	(Kim et al. 2021)
Caliciviridae	Tulane virus	Nitrogen-oxygen, 5 min	Lettuce	1.8	(Min et al. 2016)
Caliciviridae	Tulane virus	Atmospheric, 60 min	Blueberries	3.5	(Lacombe et al. 2017)
Caliciviridae	Feline calicivirus	Atmospheric, 3 min	Lettuce	> 5	(Aboubakr et al. 2015)
Adenoviridae	Adenovirus	Argon, 2.5 min	Viral suspension	None	(Bunz et al. 2018)
Leviviridae	MS2	Argon-air, 2 min	Viral suspension	> 10	(Guo et al. 2018)

1.2.5 Depuration and relaying of shellfish

A nonlethal inactivation method is important for shellfish which are typically consumed raw and processed alive. As shellfish accumulate virus particles into their digestive system, it is not possible to remove or inactivate with chemical or light-based methods. Instead, a natural filtration process is used, with the shellfish placed in clean water and allowed to be purged. This is either done by transferring the harvest to a clean estuarine site, known as relaying, or by placing them in a tank of sterile seawater, known as depuration (Campos and Lees 2014, Lees 2000). Table 8 shows some available data on the effectiveness of depuration and relaying. Due to the absence of food for the shellfish in the sterile water, depuration is carried out over time periods of less than a week. Relaying can be done over weeks or months if a clean site is available. Depuration is effective with bacterial hazards, eliminating most *Escherichia coli* levels, but much less effective for viruses like human noroviruses. The reason for this is uncertain, though it is thought to be related to both the size of the bacteria and the potential for it to bind to the digestive tissues (Le Guyader et al. 2012, Ueki et al. 2007). A recent review (McLeod et al. 2017) found that the time for a 1 log reduction was over one week, which is not commercially viable. Relaying on the other hand is almost entirely effective where the water is free from contaminants, but more cost intensive.

Table 8. The effect of depuration and relaying on reducing virus concentrations in shellfish.

Family	Virus	Treatment	Matrix	Log_{10} reduction	Reference
Reoviridae	Rotavirus	Relaying, 37 days	Oyster	< 5	(Loisy et al. 2005)
Reoviridae	Rotavirus	Depuration, 22°C, 7 days,	Oyster	1	(Loisy et al. 2005)
Reoviridae	Rotavirus	Depuration, 4 days	Mussel	< 2	(Bosch et al. 1995)
Picornaviridae	Hepatitis A virus	Depuration, 4 days	Mussel	< 2	(Bosch et al. 1995)
Picornaviridae	Poliovirus	Depuration, 4 days	Mussel	> 3	(Abad et al. 1997)
Adenoviridae	Adenovirus	Depuration, 4 days	Mussel	> 3	(Abad et al. 1997)
Caliciviridae	Norovirus	Depuration, 2 days	Oyster	None	(Schwab et al. 1998)
Caliciviridae	Norovirus	Depuration, 10 days	Oyster	None	(Ueki et al. 2007)
Caliciviridae	Feline calicivirus	Depuration, 10 days	Oyster	> 5	(Ueki et al. 2007)
Leviviridae	MS2	Relaying, 2–3 weeks	Oyster	> 5	(Humphrey and Martin 1993)
Leviviridae	MS2	Depuration, 22°C, 7 days	Oyster	2	(Loisy et al. 2005)

1.3 Food-related processing

All measures and actions taken to remove and inactivate food viruses must take into account that while the growth of viruses in food is not a problem, survival or persistence of infectivity is the key issue to be addressed. Therefore, all intrinsic and extrinsic factors of food processing technologies and chemical-based technologies could be used to control/inactivate enteric viruses from foods (Bosch et al. 2018). These factors and technological processes will influence the degree of initial contamination, the stability of the virus, and the dose of virus needed to cause an infection (Papafragkou et al. 2006).

Some of the primary factors affecting the survival of viruses in liquid environmental matrices are temperature, ionic strength, chemical component, microbial antagonism, the sorption status of the virus, and the type of virus. On inanimate surfaces (or fomites), the most important factors that affect virus stability are the type of virus and surface, relative humidity, moisture content, temperature, the composition of the suspending medium, light exposure, and presence of antiviral chemical or biological agents (Bosch et al. 2018, Sánchez and Bosch 2016).

1.3.1 Fermentation (biological processes and metabolites)

Fermented foods and beverages are defined as foods made through intentional microbial growth and enzymatic conversions of food components. For appreciable levels of fermentation produced organic acids (> 100 mM), together with low water activity, salt, nitrite and other antimicrobials, there is a long history of safety (Marco et al. 2021). These processes can be classified by the main metabolites and microorganisms e.g. alcohol and carbon dioxide (yeast), acetic acid (Acetobacter), lactic acid (lactic acid bacteria – LAB), mainly belonging to genera (*Leuconostoc, Lactobacillus* and/or *Streptococcus*), propionic acid (*Propionibacterium freudenreichii*), and ammonia and fatty acids (*Bacillus* or moulds) (Marco et al. 2017).

There are different ways of inhibiting viruses by LAB. Firstly, direct physical interaction between viruses and bacteria, and secondly stimulation of the host's immune system as well as the production of antiviral substances. The direct interaction occurs through adsorption or trapping of the virus and is strictly a strain-dependent mechanism (Al Kassaa et al. 2014). Therefore, the fermentation process or the use of selected microbial metabolites can be an important strategy for protecting food against the presence of viruses and health hazards (Raihan et al. 2021). The bio-preservative activity of LAB could be expressed by inoculation into food matrix and/or food packaging as well as a combination with other preservation methods (Arena et al. 2018). Seo et al. (2014) observed that MNV and FCV titer reductions in 5% NaCl fermented oysters, significantly decreased by 3.01 log at 15 days after fermentation. Similarly, FCV and MNV decreased about 4.12 and 1.47 log respectively after fermentation of Dongchimi (a type of fermented radish/kimchi) (Lee et al. 2012). The presence of antiviral and virucidal substances in the environment were shown in some experiments. Fujioka et al. (1980) observed 4–8 log reduction of coxsackie B-4, echovirus-7 and poliovirus type 1 after 3–4 days from the inoculation of the Hawaiian seawater. The effect was not present when seawater was sterilized. Furthermore, Deng and Cliver (1995) noticed that HAV was consistently inactivated more rapidly in the two types of mixed wastes (dairy cattle manure slurry and swine manure slurry) than in septic tank effluent alone or in the control. Whereas Toranzo et al. (1982) isolated antiviral-producing marine bacteria that had marked activity against poliovirus (net inactivation $\geq$ logs within 6–8 days); extracellular products appeared to be involved in the virus-inactivation process. Other enteric viruses were also inactivated by these marine bacteria. The research results cited are observational. Currently, the possibilities of isolating microorganisms and analyzing their metabolites are much greater than 20 years ago. Careful examination and use of this knowledge can be put into practice. Padhi et al. (2021) found that Soybean fermented using *Bacillus* spp. revealed production of specific antiviral peptides based on *in silico* analysis. Production of different antiviral peptides was dependent on the starter culture at species as well as strain level. Further analyses of the selected peptides using molecular docking studies demonstrated that two peptides could interact with the critical residues of the SARS-CoV-2 S1 receptor binding the domain and human TLR4/MD2 complex.

The rationale for such research is also the extensive use of chemicals in production: effective and environmental-friendly control methods are desirable.

1.4 Chemical processes

Viral transmissions can occur via close human-to-human contact or via contact with a contaminated surface. Thus, careful disinfection or sanitization is essential to reduce viral spread. For these purposes, virucidal agents are needed, which can either destroy viruses or alter their surface structures to prevent them from infecting potential host cells. Before applying a specific procedure, the following issues should be considered: the effective dose of each sanitizing agent; exposure time for effective virucidal activity; suitability for usage under domestic/processing or healthcare/hospital settings as well as mechanisms of action. The efficacy of disinfection agents against viruses is influenced by the contact time, concentration of disinfection agent, and the type of virus involved. Specific environmental factors such as temperature, humidity and pH also play a role (Lin et al. 2020).

The characteristics of viruses influence the range of inactivation by disinfection. Enveloped and unenveloped viruses differ in their disinfection sensitivity from the higher to the lower, respectively (Mcdonnell and Russell 1999). Enveloped viruses contain a lipid envelope that is required for infection, and therefore interfering with the envelope could potentially reduce virus infectivity. Lipophilic disinfectants can often be used to inactivate enveloped viruses. In contrast, non-enveloped viruses utilize a protein coat for infection, and therefore inactivation often requires denaturation of the redundant viral capsid proteins or essential replicative proteins (Nuanualsuwan and Cliver 2003). Therefore, it is much more challenging to inactivate small non-enveloped noroviruses, and several commonly available disinfectants are not able to sufficiently reduce infectivity (D'Souza and Su 2010). Viruses also show resistance to disinfection due to the cellular tissues that viruses are normally associated with. Viruses are dependent on host cells for replication, what causes that often are found together with cell fragments or aerosolized droplets named viral clumping protective factors, and they can reduce the effectiveness of the disinfectant to the virus (Lin et al. 2020).

Thinking about the efficacy of disinfection, attention must be paid on the cleaning processes, as the removal of organic material dirt first can allow for a better disinfection process (Gallandat et al. 2017). Another phenomenon important for disinfection efficacy is that virus can also aggregate in the environment, thereby making it more difficult for disinfectants to penetrate and access the virus (Gerba and Betancourt 2017). Table 9 summarises the virus reduction in suspension tests without organic load. Different substances used as disinfectants lead to different degrees of virus reduction. Therefore, an individual selection must be made to select the appropriate disinfectant for specific purposes (Nims and Zhou 2016).

1.4.1 Alcohol

Although formulations based on > 70% ethanol are virucidal, there is a need to reformulate products with much lower alcohol concentrations. One approach in developing virucidal formulations is to understand the mechanisms of action of active ingredients and formulation excipients (Martín-González et al. 2020).

1.4.2 Oxidizing agents

Oxidizing capability to inactivate viruses is achieved by disinfectants such as chloride and derivatives, hydrogen peroxide, and peracetic acid as well as ozone (gaseous or aqueous). Hypochlorites belong to the most widely used chlorine disinfectants, accessible as liquid (e.g., sodium hypochlorite) or solid (e.g., calcium hypochlorite). Alternative compounds that release chlorine include chlorine dioxide, sodium dichloroisocyanurate, and chloramine-T (CDC 2019). Hydrogen peroxide works by producing destructive hydroxyl free radicals that can attack membrane lipids, DNA, and other essential cell components. Catalase, produced by aerobic organisms and facultative anaerobes that possess cytochrome systems, can protect cells from metabolically produced hydrogen peroxide by degrading hydrogen peroxide to water and oxygen (CDC 2019).

1.4.3 Quaternary ammonium compounds

Quaternary ammonium compounds are widely used as disinfectants. They are virucidal against lipophilic (enveloped) viruses, but not effective against hydrophilic (nonenveloped) viruses.

1.4.4 Enzymes

Hydrolytic and oxidative enzymes can be used to deactivate pathogens, including bacteria, spores, viruses, and fungi. Laccases, haloperoxidases, and perhydrolases catalyze the generation of biocidal oxidants, such as iodine, bromine, hypohalous acid (e.g., HOCl or HOBr), and peracetic acid. These oxidants have broad-spectrum antimicrobial activity. Due to the multi-pathway action of these oxidants, it has proven extremely difficult for microbes to gain resistance (Grover et al. 2013).

Table 9. Reduction of the number of viruses in suspension tests without organic load (Lin et al. 2020).

Sanitizing agent	Advantages/Disadvantages	Concentration (%)	Virus	Exposure time	Reduction*	Temp. [°C]
Ethyl alcohol	Pros: broad-spectrum and non-staining. Cons: flammable, requires specific concentration range to be effective	70	Poliovirus (Sabin 1an)	1 min	0.4	37
		70	Murine Norovirus (CW3)	1 min	> 3.6	-
		70	Murine Norovirus (CW3)	5 min	> 3.6	-
		70	Feline Calicivirus (F9)	1 min	0.5 ± 0.6	-
		70	Feline Calicivirus (F9)	5 min	2.6 ± 0.3	-
		70	Influenza A (H1N1)	1 min	≥ 4.84	20
Isopropyl alcohol	Pros: broad-spectrum and non-staining. Cons: only inactivates lipid viruses, flammable, requires specific concentration range to be effective	70	Murine Norovirus (CW3)	1 min	2.6 ± 0.3	RT
		70	Murine Norovirus (CW3)	5 min	> 2.6	RT
		70	Feline Calicivirus (F9)	1 min	0.1 ± 0.1	RT
		70	Feline Calicivirus (F9)	5 min	0.2 ± 0.2	RT
Benzalkonium chloride (Cationic Surfactants – Quaternary Ammonium Compounds)	Odorless, colorless, and non-caustic. Requires warmer temperatures and longer reaction time. Virucidal activity reduced by the presence of contaminating organic matter.	0.2	Human Adenovirus	1 min	0.25	RT
		0.2	Poliovirus	1 min	0.12	RT
		0.2	Human Coxsackie virus	1 min	> 5.12	RT
		0.1	Human Adenovirus Type 3	1 h	5.02 ± 1.57	33
		0.1	Human Adenovirus Type 4	1 h	2.94 ± 0.80	33
Povidone-iodine (Halogenated compounds)	Long-lasting slow release of iodine, fast acting, more effective than many other disinfectants.	0.23	Influenza A subtype H1N1	15 s	5.67 ± 0.43	20
		0.023	Influenza A subtype H1N2	15 s	4.50 ± 0.54	20

	Concentration	Virus	Time	Reduction of activity	Temp (°C)
	0.23	Rotavirus strain Wa	15 s	≥4.67 ± 0.42	20
	8	Adenovirus Type 5	3 min	≥4.63	-
	8	Poliovirus Type 1	60 min	≥4.93	-
Chlorohexidine digluconate (Halogenated compounds) — Ineffective against non-enveloped viruses. Less potent and slower-acting than povidone-iodine	0.02	Murine hepatitis virus	10 min	0.7–0.8	23
	0.12	Hepatitis B virus	15 min	99%	AT
Formaldehyde — Pros: wide-spectrum activity; Cons: Pungent, hazardous to health	0.7	Murine hepatitis virus	10 min	> 3.45	23
	2.0	Human adenovirus Type 5	1 h	> 5.0	25
Sodium hypochlorite — Pros: Broad spectrum, no toxic residues, fast-acting; Cons: Corrosive to metals at high concentrations, Decreased activity in the presence of organic matter	0.016	Norwalk virus	30s	5.0	-
	0.005	Murine norovirus 3	1 min	~ 2	-
Hydrogen peroxide — Pros: Broad-spectrum, stable; Cons: Slower acting	0.1	Feline calicivirus F9 (norovirus)	15 min	> 3	20
Peracetic acid — Pros: Fast acting, leaves no residue, still effective in organic matter; Cons: corrodes metals, unstable	0.0085	Murine norovirus 3	1 min	~ 3	-

* Reduction of activity (log10) OR Reduction of virus titer (%) OR PFU/ml[x]

AT, unspecified ambient temperature; RT, room temperature

1.4.5 Plant-based antiviral natural compounds

Plant-based antimicrobial compounds are produced in various parts of the plants, e.g., flowers, buds, fruits, seeds, herbs, roots, leaves, bark, wood, and stem (Bright and Gilling 2016). Plant-based antiviral natural compounds may offer less toxicity. Common plant compounds include flavonoids, phenolics, carotenoids, terpenoids, alkaloids, and many others, many of which have documented antiviral activities (Das et al. 2021).

Antiviral effects of food components, food extracts, biochemicals and essential oils (EO) on various viruses are listed elsewhere (Bosch et al. 2018, Jama-Kmiecik et al. 2021, Sarowska et al. 2021). The results of the research were varied, from no log10 reduction to 7 log. The biggest reduction of MNV (5–7 log) was observed when carvacrol, 0.5%, Hibiscus sabdariffa extract, 40–100 mg/ml, 24 h, 37°C and grape seed extract 1–2 mg/ml, 1 h were used (Bosch et al. 2018).

1.5 Prevention strategies

Both practical (e.g., thermal/nonthermal, chemical operation and other) as well as procedural (e.g., fulfilment with guidelines and regulations) measures can help in reducing the potential risk of transmission of viruses in the food chain. Raw or only minimal processing foods are the most probable vehicles for virus transmission. Consequently, hygiene measures in the food industry are especially important for ready-to-eat food products, and they are seldom if ever subject to thorough disinfection. The most reliable control of foodborne viral contamination will therefore be to prevent the contamination from occurring in the first place, and this should be achieved by effective guidelines made available to the food industry (Cook and Richards 2013). Preharvest and postharvest control strategies mainly concern the first stage of food production (harvesting), that includes mainly shellfish, fruits and vegetables and pork. Regarding procedural measures, there are many proposals, either compulsory like GMP/GHP or HACCP in the European Union or voluntary like quality and safety management systems (e.g., BRC), which can also contribute to reducing the risk of viruses in the food and environment.

Effective usage of physical and chemical procedures in the food chain may require the use of several methods in one process. This approach comes under the hurdle technology concept. However, special attention should be paid to decontamination of hands. Virus transfer from hands to fomites and vice-versa is thought to play a significant role in the spreading of infection. Hands can be easily contaminated with norovirus by transfer from fomites, and then the contaminated hands could cross-contaminate up to 7 other surfaces without any recontamination (Alidjinou et al. 2018, Barker et al. 2004).

Other invaluable measures are education, training and supervision of all food processing workers, as well as vaccination against viruses such as HAV or rotavirus.

2. Concluding remarks

The literature on foodborne inactivation has grown significantly, as better detection methods and inactivation methods become more widely available. Established treatments like heating and gamma irradiation are becoming better understood. Nonthermal methods with lower effect on the food quality, such as HPP and cold plasma, are being proven effective in different matrices. Pulsed treatments like HIPL are also promising, although pulsed electric fields show little effect against viruses (Lee and Grove 2016). The removal of viruses entirely from foods like shellfish and fresh produce is very difficult, so the importance of prevention and sanitation is clear. To preserve the quality of foods, many different techniques can be applied together, also known as hurdle technology. To apply this technology, it is essential that it first be modelled, then validated. It requires a good understanding of individual inactivation methods and the potential of the synergistic effect. The differences in effect for the same methods, as observed in the literature, shows the importance of optimization of technique, and validation with the specific virus and food of interest. Better detection methods can provide better data, but for solid implementation in building inactivation models, it is needed that they are accessible and harmonized.

References

Abad, F.X., Pintó, R.M., Gajardo, R. and Bosch, A. 1997. Viruses in mussels: public health implications and depuration. Journal of Food Protection 60: 677–681. https://doi.org/10.4315/0362-028X-60.6.677.

Aboubakr, H.A., Williams, P., Gangal, U., Youssef, M.M., El-Sohaimy, S.A.A., Bruggeman, P.J. et al. 2015. Virucidal effect of cold atmospheric gaseous plasma on feline calicivirus, a surrogate for human norovirus. Appl. Environ. Microbiol. 81: 3612–3622. https://doi.org/10.1128/AEM.00054-15.

Ailavadi, S., Davidson, P.M., Morgan, M.T. and D'Souza, D.H. 2019. Thermal inactivation kinetics of tulane virus in cell-culture medium and spinach. Journal of Food Science 84: 557–563. https://doi.org/10.1111/1750-3841.14461.

Al Kassaa, I., Hober, D., Hamze, M., Chihib, N.E. and Drider, D. 2014. Antiviral potential of lactic acid bacteria and their bacteriocins. Probiotics & Antimicro. Prot. 6: 177–185. https://doi.org/10.1007/s12602-014-9162-6.

Alidjinou, E.K., Sane, F., Firquet, S., Lobert, P.-E. and Hober, D. 2018. Resistance of enteric viruses on fomites. Intervirology 61: 205–213. https://doi.org/10.1159/000448807.

Aman Mohammadi, M., Ahangari, H., Zabihzadeh Khajavi, M., Yousefi, M., Scholtz, V. and Hosseini, S.M. 2021. Inactivation of viruses using nonthermal plasma in viral suspensions and foodstuff: A short review of recent studies. Journal of Food Safety 41: e12919. https://doi.org/10.1111/jfs.12919.

Araud, E., DiCaprio, E., Ma, Y., Lou, F., Gao, Y., Kingsley, D. et al. 2016. Thermal inactivation of enteric viruses and bioaccumulation of enteric foodborne viruses in live oysters (Crassostrea virginica). Appl. Environ. Microbiol. 82: 2086–2099. https://doi.org/10.1128/AEM.03573-15.

Arena, M.P., Capozzi, V., Russo, P., Drider, D., Spano, G. and Fiocco, D. 2018. Immunobiosis and probiosis: antimicrobial activity of lactic acid bacteria with a focus on their antiviral and antifungal properties. Appl. Microbiol. Biotechnol. 102: 9949–9958. https://doi.org/10.1007/s00253-018-9403-9.

Baert, L., Uyttendaele, M., Van Coillie, E. and Debevere, J. 2008. The reduction of murine norovirus 1, B. fragilis HSP40 infecting phage B40-8 and E. coli after a mild thermal pasteurization process of raspberry puree. Food Microbiology 25: 871–874. https://doi.org/10.1016/j.fm.2008.06.002.

Baert, Leen, Wobus, C.E., Van Coillie, E., Thackray, L.B., Debevere, J. and Uyttendaele, M. 2008. Detection of murine norovirus 1 by using plaque assay, transfection assay, and real-time reverse transcription-PCR before and after heat exposure. Applied and Environmental Microbiology 74: 543–546. https://doi.org/10.1128/AEM.01039-07.

Baert, L., Debevere, J. and Uyttendaele, M. 2009. The efficacy of preservation methods to inactivate foodborne viruses. International Journal of Food Microbiology 131: 83–94. https://doi.org/10.1016/j.ijfoodmicro.2009.03.007.

Barker, J., Vipond, I.B. and Bloomfield, S.F. 2004. Effects of cleaning and disinfection in reducing the spread of Norovirus contamination via environmental surfaces. J. Hosp. Infect. 58: 42–49. https://doi.org/10.1016/j.jhin.2004.04.021.

Barnaud, E., Rogée, S., Garry, P., Rose, N. and Pavio, N. 2012. Thermal inactivation of infectious hepatitis E virus in experimentally contaminated food. Appl. Environ. Microbiol. 78: 5153–5159. https://doi.org/10.1128/AEM.00436-12.

Bidawid, S., Farber, J., Sattar, S.A. and Hayward, S. 2000. Heat inactivation of hepatitis A virus in dairy foods†. Journal of Food Protection 63: 522–528. https://doi.org/10.4315/0362-028X-63.4.522.

Bidawid, S., Farber, J.M. and Sattar, S.A. 2000. Inactivation of hepatitis A virus (HAV) in fruits and vegetables by gamma irradiation. International Journal of Food Microbiology 57: 91–97. https://doi.org/10.1016/S0168-1605(00)00235-X.

Bosch, A., Pinto, R.M. and Abad, F.X. 1995. Differential accumulation and depuration of human enteric viruses by mussels. Water Science and Technology, Health-Related Water Microbiology 31: 447–451. https://doi.org/10.1016/0273-1223(95)00310-J.

Bosch, A., Gkogka, E., Le Guyader, F.S., Loisy-Hamon, F., Lee, A., van Lieshout, L. et al. 2018. Foodborne viruses: Detection, risk assessment, and control options in food processing. International Journal of Food Microbiology 285: 110–128. https://doi.org/10.1016/j.ijfoodmicro.2018.06.001.

Bozkurt, H., D'Souza, D.H. and Davidson, P.M. 2014a. A comparison of the thermal inactivation kinetics of human norovirus surrogates and hepatitis A virus in buffered cell culture medium. Food Microbiology 42: 212–217. https://doi.org/10.1016/j.fm.2014.04.002.

Bozkurt, H., Leiser, S., Davidson, P.M. and D'Souza, D.H. 2014b. Thermal inactivation kinetic modeling of human norovirus surrogates in blue mussel (Mytilus edulis) homogenate. International Journal of Food Microbiology 172: 130–136. https://doi.org/10.1016/j.ijfoodmicro.2013.11.026.

Bozkurt, H., D'Souza, D.h. and Davidson, P.m. 2015a. Thermal inactivation kinetics of hepatitis A virus in homogenized clam meat (Mercenaria mercenaria). Journal of Applied Microbiology 119: 834–844. https://doi.org/10.1111/jam.12892.

Bozkurt, H., D'Souza, D.H. and Davidson, P.M. 2015b. Thermal inactivation kinetics of human norovirus surrogates and hepatitis A virus in turkey deli meat. Applied and Environmental Microbiology 81: 4850–4859. https://doi.org/10.1128/AEM.00874-15.

Bozkurt, H., Ye, X., Harte, F., D'Souza, D.H. and Davidson, P.M. 2015c. Thermal inactivation kinetics of hepatitis A virus in spinach. International Journal of Food Microbiology 193: 147–151. https://doi.org/10.1016/j.ijfoodmicro.2014.10.015.

Bradshaw, E. and Jaykus, L.-A. 2016. Risk assessment for foodborne viruses. pp. 471–503. In: Goyal, S.M. and Cannon, J.L. (eds.). Viruses in Foods, Food Microbiology and Food Safety. Springer International Publishing. https://doi.org/10.1007/978-3-319-30723-7_17.

Bright, K.R. and Gilling, D.H. 2016. Natural virucidal compounds in foods. pp. 449–469. In: Goyal, S.M. and Cannon, J.L. (eds.). Viruses in Foods, Food Microbiology and Food Safety. Springer International Publishing, Cham. https://doi.org/10.1007/978-3-319-30723-7_16.

Buckow, R., Isbarn, S., Knorr, D., Heinz, V. and Lehmacher, A. 2008. Predictive model for inactivation of feline calicivirus, a norovirus surrogate, by heat and high hydrostatic pressure. Applied and Environmental Microbiology 74: 1030–1038. https://doi.org/10.1128/AEM.01784-07.

Bunz, O., Mese, K., Zhang, W., Piwowarczyk, A. and Ehrhardt, A. 2018. Effect of cold atmospheric plasma (CAP) on human adenoviruses is adenovirus type-dependent. PLOS ONE 13: e0202352. https://doi.org/10.1371/journal.pone.0202352.

Butot, S., Putallaz, T., Amoroso, R. and Sánchez, G. 2009. Inactivation of enteric viruses in minimally processed berries and herbs. Appl. Environ. Microbiol. 75: 4155–4161. https://doi.org/10.1128/AEM.00182-09.

Calci, K.R., Meade, G.K., Tezloff, R.C. and Kingsley, D.H. 2005. High-pressure inactivation of hepatitis A virus within oysters. Applied and Environmental Microbiology 71: 339–343. https://doi.org/10.1128/AEM.71.1.339-343.2005.

Campos, C.J.A. and Lees, D.N. 2014. Environmental transmission of human noroviruses in shellfish waters. Appl. Environ. Microbiol. 80: 3552–3561. https://doi.org/10.1128/AEM.04188-13.

Cannon, J.L., Papafragkou, E., Park, G.W., Osborne, J., Jaykus, L.-A. and Vinjé, J. 2006. Surrogates for the study of norovirus stability and inactivation in the environment: a comparison of Murine norovirus and Feline Calicivirus. Journal of Food Protection 69: 2761–2765. https://doi.org/10.4315/0362-028X-69.11.2761.

CDC, C. for D.C. and P. 2019. Chemical Disinfectants | Disinfection & Sterilization Guidelines | Guidelines Library | Infection Control | CDC [WWW Document]. URL https://www.cdc.gov/infectioncontrol/guidelines/disinfection/disinfection-methods/chemical.html (accessed 2.14.22).

Chen, H., Hoover, D.G. and Kingsley, D.H. 2005. Temperature and treatment time influence high hydrostatic pressure inactivation of feline calicivirus, a norovirus surrogate†. Journal of Food Protection 68: 2389–2394. https://doi.org/10.4315/0362-028X-68.11.2389.

Choi, M.-S., Jeon, E.B., Kim, J.Y., Choi, E.H., Lim, J.S., Choi, J. et al. 2020. Virucidal effects of dielectric barrier discharge plasma on human norovirus infectivity in fresh oysters (Crassostrea gigas). Foods 9: 1731. https://doi.org/10.3390/foods9121731.

Cook, N. and Richards, G.P. 2013. An introduction to food- and waterborne viral disease. pp. 3–18. *In*: Cook, Nigel (ed.). Viruses in Food and Water, Woodhead Publishing Series in Food Science, Technology and Nutrition. Woodhead Publishing. https://doi.org/10.1533/9780857098870.1.3.

Croci, L., Ciccozzi, M., De Medici, D., Di Pasquale, S., Fiore, A., Mele, A. et al. 1999. Inactivation of Hepatitis A virus in heat-treated mussels. Journal of Applied Microbiology 87: 884–888. https://doi.org/10.1046/j.1365-2672.1999.00935.x.

Das, G., Heredia, J.B., de Lourdes Pereira, M., Coy-Barrera, E., Rodrigues Oliveira, S.M., Gutiérrez-Grijalva, E.P. et al. 2021. Korean traditional foods as antiviral and respiratory disease prevention and treatments: A detailed review. Trends in Food Science & Technology 116: 415–433. https://doi.org/10.1016/j.tifs.2021.07.037.

de Roda Husman, A.M., Bijkerk, P., Lodder, W., van den Berg, H., Pribil, W., Cabaj, A. et al. 2004. Calicivirus inactivation by nonionizing (253.7-nanometer-wavelength [UV]) and ionizing (Gamma) radiation. Applied and Environmental Microbiology 70: 5089–5093. https://doi.org/10.1128/AEM.70.9.5089-5093.2004.

Deboosere, N., Legeay, O., Caudrelier, Y. and Lange, M. 2004. Modelling effect of physical and chemical parameters on heat inactivation kinetics of hepatitis A virus in a fruit model system. International Journal of Food Microbiology 93: 73–85. https://doi.org/10.1016/j.ijfoodmicro.2003.10.015.

Deng, M.Y. and Cliver, D.O. 1995. Persistence of inoculated hepatitis A virus in mixed human and animal wastes. Applied and Environmental Microbiology. https://doi.org/10.1128/aem.61.1.87-91.1995.

DiCaprio, E., Ye, M., Chen, H. and Li, J. 2019. Inactivation of human norovirus and tulane virus by high pressure processing in simple mediums and strawberry puree. Frontiers in Sustainable Food Systems 3.

DiGirolamo, R., Liston, J. and Matches, J.R. 1970. Survival of virus in chilled, frozen, and processed oysters. Applied Microbiology 20: 58–63. https://doi.org/10.1128/am.20.1.58-63.1970.

Doultree, J.C., Druce, J.D., Birch, C.J., Bowden, D.S. and Marshall, J.A. 1999. Inactivation of feline calicivirus, a Norwalk virus surrogate. Journal of Hospital Infection 41: 51–57. https://doi.org/10.1016/S0195-6701(99)90037-3.

D'Souza, D.H. and Su, X. 2010. Efficacy of chemical treatments against murine norovirus, feline calicivirus, and MS2 bacteriophage. Foodborne Pathog. Dis. 7: 319–326. https://doi.org/10.1089/fpd.2009.0426.

Duizer, E., Bijkerk, P., Rockx, B., de Groot, A., Twisk, F. and Koopmans, M. 2004. Inactivation of caliciviruses. Applied and Environmental Microbiology 70: 4538–4543. https://doi.org/10.1128/AEM.70.8.4538-4543.2004.

Emerson, S.U., Arankalle, V.A. and Purcell, R.H. 2005. Thermal stability of hepatitis E virus. The Journal of Infectious Diseases 192: 930–933. https://doi.org/10.1086/432488.

Emmoth, E., Rovira, J., Rajkovic, A., Corcuera, E., Wilches Pérez, D., Dergel, I., Ottoson, J.R. et al. 2017. Inactivation of viruses and bacteriophages as models for swine hepatitis E virus in food matrices. Food Environ. Virol. 9: 20–34. https://doi.org/10.1007/s12560-016-9268-y.

Ettayebi, K., Crawford, S.E., Murakami, K., Broughman, J.R., Karandikar, U., Tenge, V.R. et al. 2016. Replication of human noroviruses in stem cell–derived human enteroids. Science 353: 1387–1393. https://doi.org/10.1126/science.aaf5211.

Ettayebi, K., Tenge, V.R., Cortes-Penfield, N.W., Crawford, S.E., Neill, F.H., Zeng, X.-L. et al. 2021. New Insights and enhanced human norovirus cultivation in human intestinal enteroids. mSphere 6: e01136–20. https://doi.org/10.1128/msphere.01136-20.

FAO, WHO. 2008. Viruses in Food: Scientific Advice to Support Risk Management Activities: Meeting Report. Microbiological Risk Assessment Series (MRA) 13, Microbiological Risk Assessment Series (FAO/WHO). FAO/WHO, Rome, Italy.

Feng, K., Divers, E., Ma, Y. and Li, J. 2011. Inactivation of a human norovirus surrogate, human norovirus virus-like particles, and vesicular stomatitis virus by gamma irradiation. Appl. Environ. Microbiol. 77: 3507–3517. https://doi.org/10.1128/AEM.00081-11.

Fino, V.R. and Kniel, K.E. 2008. UV light inactivation of hepatitis A virus, aichi virus, and feline calicivirus on strawberries, green onions, and lettuce. Journal of Food Protection 71: 908–913. https://doi.org/10.4315/0362-028X-71.5.908.

Foley, D.M., Dufour, A., Rodriguez, L., Caporaso, F. and Prakash, A. 2002. Reduction of *Escherichia coli* O157:H7 in shredded iceberg lettuce by chlorination and gamma irradiation. Radiation Physics and Chemistry, 12th International Meeting on Radiation Processing (IMRP-12) 63: 391–396. https://doi.org/10.1016/S0969-806X(01)00530-8.

Fujioka, R.S., Loh, P.C. and Lau, L.S. 1980. Survival of human enteroviruses in the Hawaiian ocean environment: evidence for virus-inactivating microorganisms. Applied and Environmental Microbiology.

Gallandat, K., Wolfe, M.K. and Lantagne, D. 2017. Surface cleaning and disinfection: efficacy assessment of four chlorine types using *Escherichia coli* and the ebola surrogate Phi6. Environ. Sci. Technol. 51: 4624–4631. https://doi.org/10.1021/acs.est.6b06014.

Gerba, C.P. and Betancourt, W.Q. 2017. Viral aggregation: impact on virus behavior in the environment. Environ. Sci. Technol. 51: 7318–7325. https://doi.org/10.1021/acs.est.6b05835.

Gibson, K.E. and Schwab, K.J. 2011. Thermal inactivation of human norovirus surrogates. Food Environ. Virol. 3: 74. https://doi.org/10.1007/s12560-011-9059-4.

Grove, S.F., Forsyth, S., Wan, J., Coventry, J., Cole, M., Stewart, C.M. et al. 2008. Inactivation of hepatitis A virus, poliovirus and a norovirus surrogate by high pressure processing. Innovative Food Science & Emerging Technologies, Food Innovation: Emerging Science, Technologies and Applications (FIESTA) Conference 9: 206–210. https://doi.org/10.1016/j.ifset.2007.07.006.

Grover, N., Dinu, C.Z., Kane, R.S. and Dordick, J.S. 2013. Enzyme-based formulations for decontamination: current state and perspectives. Appl. Microbiol. Biotechnol. 97: 3293–3300. https://doi.org/10.1007/s00253-013-4797-x.

Guan, D., Kniel, K., Calci, K.R., Hicks, D.T., Pivarnik, L.F. and Hoover, D.G. 2006. Response of four types of coliphages to high hydrostatic pressure. Food Microbiology 23: 546–551. https://doi.org/10.1016/j.fm.2005.09.003.

Guo, L., Xu, R., Gou, L., Liu, Z., Zhao, Y., Liu, D. et al. 2018. Mechanism of virus inactivation by cold atmospheric-pressure plasma and plasma-activated water. Appl. Environ. Microbiol. 84: e00726–18. https://doi.org/10.1128/AEM.00726-18.

Haines, J., Patel, M., Knight, A.I., Corley, D., Gibson, G., Schaaf, J. et al. 2015. Thermal inactivation of feline calicivirus in pet food processing. Food Environ. Virol. 7: 374–380. https://doi.org/10.1007/s12560-015-9211-7.

Hall, A.J., Wikswo, M.E., Pringle, K., Gould, L.H. and Parashar, U.D. 2014. Vital signs: foodborne norovirus outbreaks—United States, 2009–2012. MMWR Morb Mortal Wkly Rep 63: 491–5.

Harlow, J., Oudit, D., Hughes, A. and Mattison, K. 2011. Heat inactivation of hepatitis A virus in shellfish using steam. Food Environ. Virol. 3: 31–34. https://doi.org/10.1007/s12560-010-9052-3.

Havelaar, A.H., Kirk, M.D., Torgerson, P.R., Gibb, H.J., Hald, T., Lake, R.J. et al. 2010. Group on behalf of W.H.O.F.D.B.E.R. 2015. World Health Organization Global Estimates and Regional Comparisons of the Burden of Foodborne Disease in 2010. PLOS Med. 12: e1001923. https://doi.org/10.1371/journal.pmed.1001923.

Hewitt, J. and Greening, G.E. 2006. Effect of heat treatment on hepatitis A virus and norovirus in new zealand greenshell mussels (Perna canaliculus) by quantitative real-time reverse transcription PCR and cell culture. Journal of Food Protection 69: 2217–2223. https://doi.org/10.4315/0362-028X-69.9.2217.

Hewitt, J., Rivera-Aban, M. and Greening, G.e. 2009. Evaluation of murine norovirus as a surrogate for human norovirus and hepatitis A virus in heat inactivation studies. Journal of Applied Microbiology 107: 65–71. https://doi.org/10.1111/j.1365-2672.2009.04179.x.

Hirneisen, K.A., Black, E.P., Cascarino, J.L., Fino, V.R., Hoover, D.G. and Kniel, K.E. 2010. Viral inactivation in foods: a review of traditional and novel food-processing technologies. Comprehensive Reviews in Food Science and Food Safety 9: 3–20. https://doi.org/10.1111/j.1541-4337.2009.00092.x.

Hirneisen, K.A. and Kniel, K.E. 2013. Inactivation of internalized and surface contaminated enteric viruses in green onions. International Journal of Food Microbiology 166: 201–206. https://doi.org/10.1016/j.ijfoodmicro.2013.07.013.

Hoover, D.G., Metrick, C., Papineau, A.M., Farkas, D.F. and Knorr, D. 1989. Biological effects of high hydrostatic pressure on food microorganisms. Food Technology (USA).

Huang, Y. and Chen, H. 2015. Inactivation of *Escherichia coli* O157:H7, Salmonella and human norovirus surrogate on artificially contaminated strawberries and raspberries by water-assisted pulsed light treatment. Food Research International 72: 1–7. https://doi.org/10.1016/j.foodres.2015.03.013.

Huang, Y., Ye, M., Cao, X. and Chen, H. 2017. Pulsed light inactivation of murine norovirus, Tulane virus, *Escherichia coli* O157:H7 and Salmonella in suspension and on berry surfaces. Food Microbiology 61: 1–4. https://doi.org/10.1016/j.fm.2016.08.001.

Huang, Y.-M., Chang, W.-C. and Hsu, C.-L. 2021. Inactivation of norovirus by atmospheric pressure plasma jet on salmon sashimi. Food Res. Int. 141: 110108. https://doi.org/10.1016/j.foodres.2021.110108.

Humphrey, T.J. and Martin, K. 1993. Bacteriophage as models for virus removal from Pacific oysters (Crassostrea gigas) during re-laying. Epidemiology & Infection 111: 325–335. https://doi.org/10.1017/S0950268800057034.

Imagawa, T., Sugiyama, R., Shiota, T., Li, T.-C., Yoshizaki, S., Wakita, T. et al. 2018. Evaluation of heating conditions for inactivation of hepatitis E virus genotypes 3 and 4. Journal of Food Protection 81: 947–952. https://doi.org/10.4315/0362-028X.JFP-17-290.

Imamura, S., Kanezashi, H., Goshima, T., Suto, A., Ueki, Y., Sugawara, N. et al. 2017. Effect of high-pressure processing on human noroviruses in laboratory-contaminated oysters by bio-accumulation. Foodborne Pathogens and Disease 14: 518–523. https://doi.org/10.1089/fpd.2017.2294.

Jama-Kmiecik, A., Sarowska, J., Wojnicz, D., Choroszy-Król, I. and Frej-Mądrzak, M. 2021. Natural products and their potential anti-HAV activity. Pathogens 10: 1095. https://doi.org/10.3390/pathogens10091095.

Jean, J., Morales-Rayas, R., Anoman, M.-N. and Lamhoujeb, S. 2011. Inactivation of hepatitis A virus and norovirus surrogate in suspension and on food-contact surfaces using pulsed UV light (pulsed light inactivation of food-borne viruses). Food Microbiol. 28: 568–572. https://doi.org/10.1016/j.fm.2010.11.012.

Jung, P.-M., Park, J.S., Park, J.-G., Park, J.-N., Han, I.-J., Song, B.-S. et al. 2009. Radiation sensitivity of poliovirus, a model for norovirus, inoculated in oyster (Crassostrea gigas) and culture broth under different conditions. Radiation Physics and Chemistry, 15th International Meeting on Radiation Processing 78: 597–599. https://doi.org/10.1016/j.radphyschem.2009.03.017.

Khadre, M.A. and Yousef, A.E. 2002. Susceptibility of human rotavirus to ozone, high pressure, and pulsed electric field. Journal of Food Protection 65: 1441–1446. https://doi.org/10.4315/0362-028X-65.9.1441.

Kim, J.Y., Jeon, E.B., Choi, M.-S. and Park, S.Y. 2021. Inactivation of Human Norovirus GII. 4 on Oyster Crassostrea gigas by electron beam irradiation. Korean Journal of Fisheries and Aquatic Sciences 54: 16–22. https://doi.org/10.5657/KFAS.2021.0016.

Kingsley, D.H., Hoover, D.G., Papafragkou, E. and Richards, G.P. 2002. Inactivation of hepatitis A virus and a calicivirus by high hydrostatic pressure†. Journal of Food Protection 65: 1605–1609. https://doi.org/10.4315/0362-028X-65.10.1605.

Kingsley, D.H., Chen, H. and Hoover, D.G. 2004. Inactivation of selected picornaviruses by high hydrostatic pressure. Virus Research 102: 221–224. https://doi.org/10.1016/j.virusres.2004.01.030.

Kingsley, D.H., Guan, D. and Hoover, D.G. 2005. Pressure inactivation of hepatitis a virus in strawberry puree and sliced green onions†. Journal of Food Protection 68: 1748–1751. https://doi.org/10.4315/0362-028X-68.8.1748.

Kingsley, D.H., Holliman, D.R., Calci, K.R., Chen, H. and Flick, G.J. 2007. Inactivation of a norovirus by high-pressure processing. Applied and Environmental Microbiology 73: 581–585. https://doi.org/10.1128/AEM.02117-06.

Kingsley, D.H. and Chen, H. 2009. Influence of pH, salt, and temperature on pressure inactivation of hepatitis A virus. International Journal of Food Microbiology 130: 61–64. https://doi.org/10.1016/j.ijfoodmicro.2009.01.004.

Knight, Angus, Haines, J., Stals, A., Li, D., Uyttendaele, M., Knight, Alastair and Jaykus, L.-A. 2016. A systematic review of human norovirus survival reveals a greater persistence of human norovirus RT-qPCR signals compared to those of cultivable surrogate viruses. International Journal of Food Microbiology 216: 40–49. https://doi.org/10.1016/j.ijfoodmicro.2015.08.015.

Lacombe, A., Niemira, B.A., Gurtler, J.B., Sites, J., Boyd, G., Kingsley, D.H. et al. 2017. Nonthermal inactivation of norovirus surrogates on blueberries using atmospheric cold plasma. Food Microbiol. 63: 1–5. https://doi.org/10.1016/j.fm.2016.10.030.

Laird, D.T., Sun, Y., Reineke, K.F. and Carol Shieh, Y. 2011. Effective hepatitis A virus inactivation during low-heat dehydration of contaminated green onions. Food Microbiology 28: 998–1002. https://doi.org/10.1016/j.fm.2011.01.011.

Lamont, Y., Rzeżutka, A., Anderson, J.g., MacGregor, S.j., Given, M.j., Deppe, C. et al. 2007. Pulsed UV-light inactivation of poliovirus and adenovirus. Letters in Applied Microbiology 45: 564–567. https://doi.org/10.1111/j.1472-765X.2007.02261.x.

Le Guyader, F.S., Atmar, R.L. and Le Pendu, J. 2012. Transmission of viruses through shellfish: when specific ligands come into play. Curr. Opin. Virol., Virus entry/Environmental Virology 2: 103–110. https://doi.org/10.1016/j.coviro.2011.10.029.

Lee, A. and Grove, S. 2016. Virus inactivation during food processing. pp. 421–447. In: Goyal, S.M. and Cannon, J.L. (eds.). Viruses in Foods. Springer International Publishing, Cham. https://doi.org/10.1007/978-3-319-30723-7_1.

Lee, M.H., Yoo, S.-H., Ha, S.-D. and Choi, C. 2012. Inactivation of feline calicivirus and murine norovirus during Dongchimi fermentation. Food Microbiology 31: 210–214. https://doi.org/10.1016/j.fm.2012.04.002.

Lees, D. 2000. Viruses and bivalve shellfish. International Journal of Food Microbiology 59: 81–116.

Li, D., Tang, Q.-J., Wang, J.-F., Wang, Y.-M., Zhao, Q. and Xue, C.-H. 2009. Effects of high-pressure processing on murine norovirus-1 in oysters (Crassostrea gigas) in situ. Food Control 20: 992–996. https://doi.org/10.1016/j.foodcont.2008.11.012.

Li, X., Chen, H. and Kingsley, D.H. 2013. The influence of temperature, pH, and water immersion on the high hydrostatic pressure inactivation of GI.1 and GII.4 human noroviruses. International Journal of Food Microbiology 167: 138–143. https://doi.org/10.1016/j.ijfoodmicro.2013.08.020.

Li, X., Huang, R. and Chen, H. 2017. Evaluation of assays to quantify infectious human norovirus for heat and high-pressure inactivation studies using tulane virus. Food Environ. Virol. 9: 314–325. https://doi.org/10.1007/s12560-017-9288-2.

Lin, Q., Lim, J.Y.C., Xue, K., Yew, P.Y.M., Owh, C., Chee, P.L. and Loh, X.J. 2020. Sanitizing agents for virus inactivation and disinfection. View (Beijing) e16. https://doi.org/10.1002/viw2.16.

Liu, C., Li, X. and Chen, H. 2015. Application of water-assisted ultraviolet light processing on the inactivation of murine norovirus on blueberries. International Journal of Food Microbiology 214: 18–23. https://doi.org/10.1016/j.ijfoodmicro.2015.07.023.

Loisy, F., Atmar, R.L., Le Saux, J.-C., Cohen, J., Caprais, M.-P., Pommepuy, M. et al. 2005. Use of rotavirus virus-like particles as surrogates to evaluate virus persistence in shellfish. Applied and Environmental Microbiology 71: 6049–6053. https://doi.org/10.1128/AEM.71.10.6049-6053.2005.

Lou, F., Neetoo, H., Chen, H. and Li, J. 2011. Inactivation of a human norovirus surrogate by high-pressure processing: effectiveness, mechanism, and potential application in the fresh produce industry. Applied and Environmental Microbiology 77: 1862–1871. https://doi.org/10.1128/AEM.01918-10.

Lytle, C.D. and Sagripanti, J.-L. 2005. Predicted inactivation of viruses of relevance to biodefense by solar radiation. Journal of Virology 79: 14244–14252. https://doi.org/10.1128/JVI.79.22.14244-14252.2005.

Mallett, J.C., Beghian, L.E., Metcalf, T.G. and Kaylor, J.D. 1991. Potential of irradiation technology for improved shellfish sanitation. Journal of Food Safety 11: 231–245. https://doi.org/10.1111/j.1745-4565.1991.tb00055.x.

Marco, M.L., Heeney, D., Binda, S., Cifelli, C.J., Cotter, P.D., Foligné, B. et al. 2017. Health benefits of fermented foods: microbiota and beyond. Current Opinion in Biotechnology, Food Biotechnology • Plant Biotechnology 44: 94–102. https://doi.org/10.1016/j.copbio.2016.11.010.

Marco, M.L., Sanders, M.E., Gänzle, M., Arrieta, M.C., Cotter, P.D., De Vuyst, L. et al. 2021. The International Scientific Association for Probiotics and Prebiotics (ISAPP) consensus statement on fermented foods. Nature Reviews Gastroenterology & Hepatology 18: 196–208. https://doi.org/10.1038/s41575-020-00390-5.

Martín-González, N., Vieira Gonçalves, L., Condezo, G.N., San Martín, C., Rubiano, M., Fallis, I. et al. 2020. Virucidal action mechanism of alcohol and divalent cations against human adenovirus. Frontiers in Molecular Biosciences 7.

Mcdonnell, G. and Russell, A.D. 1999. Antiseptics and disinfectants: activity, action, and resistance. Clin. Microbiol. Rev. 12: 33.

McLeod, C., Polo, D., Saux, J.-C.L. and Guyader, F.S.L. 2017. Depuration and relaying: a review on potential removal of norovirus from oysters. Comprehensive Reviews in Food Science and Food Safety 16: 692–706. https://doi.org/10.1111/1541-4337.12271.

Min, S.C., Roh, S.H., Niemira, B.A., Sites, J.E., Boyd, G. and Lacombe, A. 2016. Dielectric barrier discharge atmospheric cold plasma inhibits *Escherichia coli* O157:H7, Salmonella, *Listeria monocytogenes*, and Tulane virus in Romaine lettuce. Int. J. Food Microbiol. 237: 114–120. https://doi.org/10.1016/j.ijfoodmicro.2016.08.025.

Molteni, M. and Donazzi, A. 2020. Model analysis of atmospheric non-thermal plasma for methane abatement in a gas phase dielectric barrier discharge reactor. Chemical Engineering Science 212. https://doi.org/10.1016/j.ces.2019.115340.

Moon, Y., Han, S., Son, J. won, Park, S.H. and Ha, S.-D. 2021. Impact of ultraviolet-C and peroxyacetic acid against murine norovirus on stainless steel and lettuce. Food Control 130: 108378. https://doi.org/10.1016/j.foodcont.2021.108378.

Murchie, L.W., Kelly, A.L., Wiley, M., Adair, B.M. and Patterson, M. 2007. Inactivation of a calicivirus and enterovirus in shellfish by high pressure. Innovative Food Science & Emerging Technologies 8: 213–217. https://doi.org/10.1016/j.ifset.2006.11.003.

Nasheri, N., Doctor, T., Chen, A., Harlow, J. and Gill, A. 2020. Evaluation of high-pressure processing in inactivation of the hepatitis E virus. Frontiers in Microbiology 11.

Nims, R.W. and Zhou, S.S. 2016. Intra-family differences in efficacy of inactivation of small, non-enveloped viruses. Biologicals 44: 456–462. https://doi.org/10.1016/j.biologicals.2016.05.005.

Nuanualsuwan, S. and Cliver, D.O. 2003. Capsid functions of inactivated human picornaviruses and feline calicivirus. Appl. Environ. Microbiol. 69: 350–357. https://doi.org/10.1128/AEM.69.1.350-357.2003.

O' Mahony, J., O' Donoghue, M., Morgan, J.G. and Hill, C. 2000. Rotavirus survival and stability in foods as determined by an optimised plaque assay procedure. International Journal of Food Microbiology 61: 177–185. https://doi.org/10.1016/S0168-1605(00)00378-0.

Ortiz-Solà, J., Viñas, I., Aguiló-Aguayo, I., Bobo, G. and Abadias, M. 2021. An innovative water-assisted UV-C disinfection system to improve the safety of strawberries frozen under cryogenic conditions. Innovative Food Science & Emerging Technologies 73: 102756. https://doi.org/10.1016/j.ifset.2021.102756.

Padhi, S., Sanjukta, S., Chourasia, R., Labala, R.K., Singh, S.P. and Rai, A.K. 2021. A multifunctional peptide from bacillus fermented soybean for effective inhibition of SARS-CoV-2 S1 receptor binding domain and modulation of toll like receptor 4: a molecular docking study. Frontiers in Molecular Biosciences 8.

Papafragkou, E., D'Souza, D.H. and Jaykus, L.-A. 2006. Food-borne viruses: prevention and control. pp. 289–330. *In*: Goyal, S.M. (ed.). Viruses in Foods. Springer US, Boston, MA. https://doi.org/10.1007/0-387-29251-9_13.

Park, G.W., Linden, K.G. and Sobsey, M.D. 2011. Inactivation of murine norovirus, feline calicivirus and echovirus 12 as surrogates for human norovirus (NoV) and coliphage (F+) MS2 by ultraviolet light (254 nm) and the effect of cell association on UV inactivation. Lett. Appl. Microbiol. 52: 162–167. https://doi.org/10.1111/j.1472-765X.2010.02982.x.

Park, S.Y., Kim, A.-N., Lee, K.-H. and Ha, S.-D. 2015. Ultraviolet-C efficacy against a norovirus surrogate and hepatitis A virus on a stainless steel surface. International Journal of Food Microbiology 211: 73–78. https://doi.org/10.1016/j.ijfoodmicro.2015.07.006.

Parry, J.V. and Mortimer, P.P. 1984. The heat sensitivity of hepatitis A virus determined by a simple tissue culture method. Journal of Medical Virology 14: 277–283. https://doi.org/10.1002/jmv.1890140312.

Praveen, C., Dancho, B.A., Kingsley, D.H., Calci, K.R., Meade, G.K., Mena, K.D. et al. 2013. Susceptibility of murine norovirus and hepatitis A virus to electron beam irradiation in oysters and quantifying the reduction in potential infection risks. Appl. Environ. Microbiol. 79: 3796–3801. https://doi.org/10.1128/AEM.00347-13.

Predmore, A., Sanglay, G.C., DiCaprio, E., Li, J., Uribe, R.M. and Lee, K. 2015. Electron beam inactivation of Tulane virus on fresh produce, and mechanism of inactivation of human norovirus surrogates by electron beam irradiation. International Journal of Food Microbiology 198: 28–36. https://doi.org/10.1016/j.ijfoodmicro.2014.12.024.

Raihan, T., Rabbee, M.F., Roy, P., Choudhury, S., Baek, K.-H. and Azad, A.K. 2021. Microbial metabolites: the emerging hotspot of antiviral compounds as potential candidates to avert viral pandemic alike COVID-19. Frontiers in Molecular Biosciences 8.

Roberts, P. and Hope, A. 2003. Virus inactivation by high intensity broad spectrum pulsed light. Journal of Virological Methods 110: 61–65. https://doi.org/10.1016/S0166-0934(03)00098-3.

Sánchez, G. and Bosch, A. 2016. Survival of enteric viruses in the environment and food. Viruses in Foods 367–392. https://doi.org/10.1007/978-3-319-30723-7_13.

Sarowska, J., Wojnicz, D., Jama-Kmiecik, A., Frej-Mądrzak, M. and Choroszy-Król, I. 2021. Antiviral potential of plants against noroviruses. Molecules 26: 4669. https://doi.org/10.3390/molecules26154669.

Sauerbrei, A. and Wutzler, P. 2009. Testing thermal resistance of viruses. Arch. Virol. 154: 115–119. https://doi.org/10.1007/s00705-008-0264-x.

Schwab, K.J., Neill, F.H., Estes, M.K., Metcalf, T.G. and Atmar, R.L. 1998. Distribution of norwalk virus within shellfish following bioaccumulation and subsequent depuration by detection using RT-PCR. J. Food Prot. 61: 1674–1680. https://doi.org/10.4315/0362-028X-61.12.1674.

Seo, D.J., Lee, M.H., Seo, J., Ha, S.-D. and Choi, C. 2014. Inactivation of murine norovirus and feline calicivirus during oyster fermentation. Food Microbiology 44: 81–86. https://doi.org/10.1016/j.fm.2014.05.016.

Shao, L., Chen, H., Hicks, D. and Wu, C. 2018. Thermal inactivation of human norovirus surrogates in oyster homogenate. International Journal of Food Microbiology 281: 47–53. https://doi.org/10.1016/j.ijfoodmicro.2018.05.013.

Sharma, M., Shearer, A.E.H., Hoover, D.G., Liu, M.N., Solomon, M.B. and Kniel, K.E. 2008. Comparison of hydrostatic and hydrodynamic pressure to inactivate foodborne viruses. Innovative Food Science & Emerging Technologies 9: 418–422. https://doi.org/10.1016/j.ifset.2008.05.001.

Slomka, M.J. and Appleton, H. 1998. Feline calicivirus as a model system for heat inactivation studies of small round structured viruses in shellfish. Epidemiology & Infection 121: 401–407. https://doi.org/10.1017/S0950268898001290.

Sow, H., Desbiens, M., Morales-Rayas, R., Ngazoa, S.E. and Jean, J. 2011. Heat inactivation of hepatitis A virus and a norovirus surrogate in soft-shell clams (Mya arenaria). Foodborne Pathogens and Disease 8: 387–393. https://doi.org/10.1089/fpd.2010.0681.

Strazynski, M., Krämer, J. and Becker, B. 2002. Thermal inactivation of poliovirus type 1 in water, milk and yoghurt. International Journal of Food Microbiology 74: 73–78. https://doi.org/10.1016/S0168-1605(01)00708-5.

Sullivan, R., Scarpino, P.V., Fassolitis, A.C., Larkin, E.P. and Peeler, J.T. 1973. Gamma radiation inactivation of coxsackievirus B-2. Applied Microbiology 26: 14–17. https://doi.org/10.1128/am.26.1.14-17.1973.

Topping, J.R., Schnerr, H., Haines, J., Scott, M., Carter, M.J., Willcocks, M.M. et al. 2009. Temperature inactivation of Feline calicivirus vaccine strain FCV F-9 in comparison with human noroviruses using an RNA exposure assay and reverse transcribed quantitative real-time polymerase chain reaction—A novel method for predicting virus infectivity. Journal of Virological Methods 156: 89–95. https://doi.org/10.1016/j.jviromet.2008.10.024.

Toranzo, A.E., Barja, J.L. and Hetrick, F.M. 1982. Antiviral activity of antibiotic-producing marine bacteria. Can J. Microbiol. 28: 231–238. https://doi.org/10.1139/m82-031.

Ueki, Y., Shoji, M., Suto, A., Tanabe, T., Okimura, Y., Kikuchi, Y. et al. 2007. Persistence of caliciviruses in artificially contaminated oysters during depuration. Applied and Environmental Microbiology 73: 5698–5701. https://doi.org/10.1128/AEM.00290-07.

Vimont, A., Fliss, I. and Jean, J. 2014. Inactivation of foodborne viruses. pp. 471–495. *In*: Bhat, R. and Gómez-López, V.M. (eds.). Practical Food Safety. John Wiley & Sons, Ltd.

Wilkinson, N., Kurdziel, A.S., Langton, S., Needs, E. and Cook, N. 2001. Resistance of poliovirus to inactivation by high hydrostatic pressures. Innovative Food Science & Emerging Technologies 2: 95–98. https://doi.org/10.1016/S1466-8564(01)00035-2.

Xie, Y., Hajdok, C., Mittal, G.S. and Warriner, K. 2008. Inactivation of MS2 F(+) coliphage on lettuce by a combination of UV light and hydrogen peroxide. Journal of Food Protection 71: 903–907. https://doi.org/10.4315/0362-028X-71.5.903.

Chapter 9

Quantitative Virus Risk Assessment in Food, Water and the Environment

Kevin Hunt,[1] *Monika Trzaskowska*[2] and *David Rodríguez-Lázaro*[3,4,]*

‖‖

1. Introduction

For centuries, people have used mathematics to analyze and predict the uncertainties of the world. This was especially true for the uncertainties of planning or undertaking risky ventures. Over time, these considerations were formalized in mathematical theories and theorems. Key to the development of risk analysis was Pascal's formulation of probability theory in the 17th century, applied to varied topics such as life expectancy, the influence of the moon on health and the annual number of Prussian soldiers dying from horsekicks. Another crucial component in the early development of risk analysis was the use of scientific methods for assessing the causal links between adverse health effects and hazardous substances and activities (Covello and Mumpower 1985).

Gradually, risk analysis and risk assessment began to be applied in several diverse fields, from toxicology and ecology, to engineering and economics. A formal process for estimating the risk of chemical substances was first introduced in US federal agencies in the late 1970s, as a means of standardizing the basis for regulatory decision making (Lammerding 1997). The difficulty of predicting an uncertain

---
[1] Centre for Food Safety, School of Biosystems and Food Engineering, University College Dublin, Belfield, Dublin 4, Ireland.
[2] Institute of Human Nutrition Sciences, Department of Food Gastronomy and Food Hygiene, Warsaw University of Life Sciences–SGGW, Nowoursynowska 159c Str., 02-776 Warsaw, Poland.
[3] Microbiology Division. Department of Biotechnology and Food Science, Faculty of Sciences, University of Burgos, Plaza Misael Bañuelos s/n, 09001 Burgos, Spain.
[4] Research Centre for Emerging Pathogens and Global Health. University of Burgos, Burgos, Spain.
* Corresponding author: drlazaro@ubu.es

future is what drives the need for risk assessment. However, there are some trends in risk analysis that can be expected to continue. Public concern about risks will rise concordant with the changing nature of social problems. New threats from emerging technologies or natural hazards will increase public perception of risk, due to the uncertain and potentially disastrous nature of perceived unforeseen consequences.

The bovine spongiform encephalopathy (BSE) crisis of the 1990s is an example of new technological hazards whose consequences were disastrous. This and other food outbreaks initiated intense international discussion on measures to prevent food-related risks and critical situations. The Agreement on the Application of Sanitary and Phytosanitary Measures (SPS Agreement) signed in 1995 during the Uruguay Round of the World Trade Organization (WTO), recognized that for the establishment of rational harmonized regulation and standards for food in international trade, a rigorous scientific process is required. The risk assessment was identified as an appropriate tool to cope with this issue. Codex Alimentarius Commission standards, guidelines and recommendations were developed to reflect international consensus regarding the requirements to protect human health from foodborne hazards. Subsequently, FAO/WHO experts discussed and formulated definitions and guidelines on how to implement this recommendation on food standards (FAO and WHO 1995). Among the documents concerning risk analysis and risk assessment in food and feed, elaborated and special attention was paid to microbial risk assessment (MRA). Three technical guidance documents were eventually published in the Microbiological Risk Assessment Series: "Hazard Characterization for Pathogens in Food and Water" in 2003 (FAO and WHO 2003), "Exposure Assessment of Microbiological Hazards in Food" in 2008 (FAO and WHO 2008a), and "Risk Characterization of Microbiological Hazards in Food" in 2009 (FAO and WHO 2009). To consolidate and update the existing technical guidance documents on microbiological risk assessment, FAO and WHO established a group of experts and convened an Expert Meeting in Rome, Italy on 11–15 March 2019. The discussions and conclusion of this meeting were taken into consideration while finalizing the report titled "Microbiological Risk Assessment—Guidance for Food" which was released in 2021 (FAO and WHO 2021).

FAO, WHO and Codex standards and related texts are voluntary. They need to be incorporated into national legislation in order to be enforceable. An example of this translation is the European Union (EU) food law regulations. In particular, Regulation (EC) No. 178/2002 of the European Parliament and of the Council of 28 January 2002, laying down the general principles and requirements of food law, establishing the European Food Safety Authority and laying down procedures in matters of food safety [OJ L 31, 1.2.2002, p. 1–24]. One of the general principles of food and feed law is the principle of risk analysis as defined by CAC/FAO/WHO.

1.1 Risk assessment

The safety of food supply must be based on solid scientific evidence and on the development of processes and procedures that utilize the available science in a rational way to arrive at public policy decisions. The importance of risk assessment lies not only in its ability to estimate human risk, but also in its use as a framework for organizing data as well as for allocating responsibility for analysis. It is important to understand that risk assessment is a process that can include a variety of models to reach a conclusion. FAO and WHO distinguish risk assessment from risk management and communication but recognized that risk assessment and risk management have a number of significant interfaces. For example, establishing priorities and policies for risk assessment often includes input from risk management consideration (FAO and WHO 1995).

According to the Codex Alimentarius Commission guidelines, risk is "*a function of the probability of an adverse health effect and the severity of that effect, consequential to hazard(s) in food*" (FAO and WHO 2019). Here a hazard is a biological, chemical or physical agent in food, or a condition of the food itself, with the potential to cause an adverse health effect. Describing risks and hazards in detail is called risk analysis. That is, "*a process consisting of three components: risk assessment, risk management and risk communication*" (FAO and WHO 2019). The first component—risk assessment—relies on a "*scientifically based process consisting of the following steps: hazard identification, hazard characterization, exposure assessment, and risk characterization*" (FAO and WHO 2019) (Figure 1). Hazard characterization, exposure assessment, and risk characterization can be estimated in a qualitative and/or quantitative manner. In brief, qualitative assessment generates textual data (non-numerical). Quantitative assessment yields numerical data, or information that can be converted into numbers.

The objective of the risk assessment process is fundamentally to answer three risk questions: "What can go wrong?", "How likely is that to happen?", and "What would the consequences be if it did go wrong?" (Lammerding 1997).

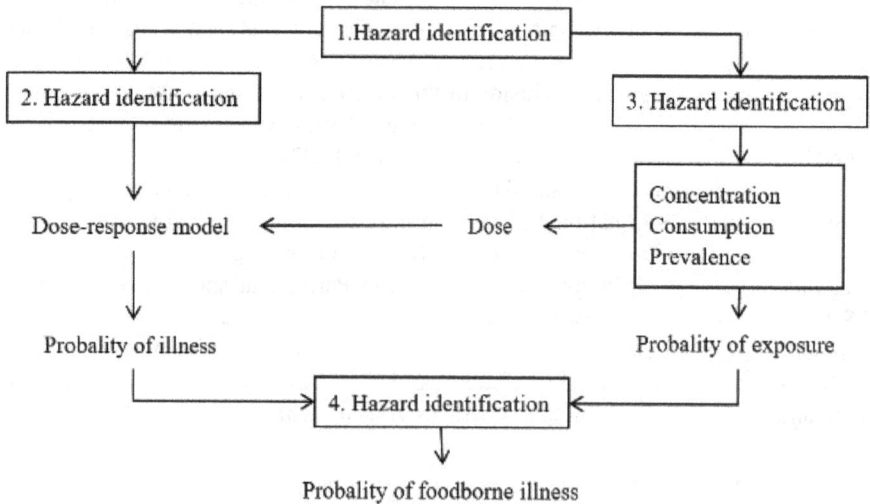

Figure 1. Flow diagram of four-step risk assessment (Ruchusatsawat et al. 2021).

1.2 Microbial risk assessment

Microbial Risk Assessment (MRA) is focused on biological hazards, mainly pathogenic bacteria. However, adverse health effects of other biological agents such as foodborne viruses, parasites are increasingly considered for assessment. General principles of best-practice risk assessment elaborated by Codex Alimentarius Commission (CAC 1999) are presented in Table 1. Most often, risk assessment is undertaken to support risk managers in their actions. Prior to the start of a risk assessment, the purpose and scope should be clearly defined. For example, by formulating appropriate risk questions, and it could also be a useful discussion among all relevant parties, including the risk managers, risk assessment team, risk communication specialist and relevant stakeholders. The scope may need to be revised during the preparation of the risk assessment if it becomes evident that the original scope cannot be achieved. Any change in scope should be discussed and agreed with the risk manager (FAO and WHO 2021).

Table 1. General principles of best-practice risk assessment (CAC 1999).

Microbiological Risk Assessment should be based upon science.
There should be a functional separation between Risk Assessment and Risk Management.
Microbiological Risk Assessment should be conducted according to a structured approach that includes Hazard Identification, Hazard Characterization, Exposure Assessment, and Risk Characterization.
Microbiological Risk Assessment should clearly state the purpose of the exercise, including the form of Risk Estimate that will be the output.
Microbiological Risk Assessment should be conducted transparently.
Any constraints that impact the Risk Assessment, such as cost, resources or time, should be identified and their possible consequences described.
Risk Estimate should contain a description of uncertainty and where the uncertainty arose during the Risk Assessment process.
Data should be such that uncertainty in the Risk Estimate can be determined. Data and data collection systems should, as far as possible, be of sufficient quality and precision so that uncertainty in the Risk Estimate is minimized.
Microbiological Risk Assessment should explicitly consider the dynamics of microbiological growth, survival, and death in foods and the complexity of the interaction (including sequelae) between human and agent following consumption as well as the potential for further spread.
Wherever possible, Risk Estimates should be reassessed over time after comparison with independent human illness data.
Microbiological Risk Assessment may need re-evaluation, as new relevant information becomes available.

1.2.1 Hazard identification

According to Codex guidelines, hazard identification is defined as "*the identification of biological, chemical and physical agents capable of causing adverse health effects and which may be present in a particular food or group of foods*" (CAC 1999). When the risk question and/or the aim and scope of risk assessment is defined, hazard identification is performed. It serves to establish the hazard as likely or real in the food product. The documentation of important information about the relationships and interactions between the hazard, the food, and the host is specified, as well as their relationship to human illness (FAO and WHO 2021). During this step of risk assessment, sensitive populations are identified, along with highlighting severity and acuteness of the illness and other complications such as long-term sequelae. The main types of data sources providing useful information to the hazard identification process are epidemiologic and clinical research data. Regarding this stage, one must remember that hazard identification may provide only a general overview, while the later steps of the risk assessment document more detailed information, such as the extent of exposure to the hazard or the dose–response relationship (FAO and WHO 2021).

1.2.2 Exposure assessment

Similarly, the exposure assessment can be defined, according to the Codex guidelines, as the qualitative and/or quantitative evaluation of the likely intake of biological, chemical, and physical agents via food as well as exposures from other sources if relevant (CAC 1999). Based on the definition of the exposure assessment, the stage is often specific to the production, processing and consumption patterns within a country or region. In addition, the activities are iterative, as repeated discussion among involved parties may lead to the refinement of the initial question to be addressed. The exposure assessment usually provides an estimate of the level of exposure to a hazard in a given population. Depending on the information available, the probability and magnitude of exposure to the hazard is calculated (FAO and WHO 2021). During this stage, the factors that have a direct effect on consumer exposure to the hazard are considered, such as frequency of consumption of the product or commodity; frequency and levels of contamination with the hazard; and factors that affect the concentration. For example, potential for microbial growth, inactivation during cooking and/or other processes, meal size, seasonal and regional effects. The other task of exposure assessment is to describe the relevant pathways of exposure. It could be the entire production-to-consumption pathway or only the pathways from retail to consumers, depending on the risk assessment purpose. The level of detail required in the different pathways reflects the question asked and the information needed by the risk managers and may be modified based on the information (FAO and WHO 2021). Detailed exposure data, characterizing the extent of microbiological hazards present in foods at the time of consumption, are usually not available. Thus, exposure assessment will often rely on a model, encompassing knowledge of the factors and their interactions that affect the number and distribution of the hazard in foods, to estimate exposure at consumption (FAO and WHO 2021).

1.2.3 Hazard characterization

Hazard characterization can be defined as the qualitative and/or quantitative evaluation of the nature of the adverse health effects associated with biological, chemical and physical agents, which may be present in food (CAC 1999). It aims to establish, for the population of interest, the probability of an adverse health effect as a function of dose. This would ideally take the form of a dose-response relationship if available, or one based on other indicators. The hazard characterization for a specific hazard can serve as a common module or component of a risk assessment performed for different purposes and within a range of goods. In general, the hazard characterization is transferable between risk assessments for the same hazard in a different context. This is because the host response to a particular hazard is not considered to be based on geography or culture (FAO and WHO 2021). Descriptive hazard characterization is used to organize and present the available information on the spectrum of human diseases associated with a particular hazard and how it is influenced by the host characteristics, the hazard and the matrix. This is based on a qualitative or semi-quantitative analysis of the available evidence and takes into account different disease mechanisms (FAO and WHO 2021).

1.2.4 Risk characterization

Finally, risk characterization is defined as the process of determining the qualitative and/or quantitative estimation, including attendant uncertainties, of the probability of occurrence and severity of known or potential adverse health effects in a given population based on hazard identification, hazard characterization and exposure assessment (CAC 1999). The risk characterization integrates the findings from previous components to estimate levels of risk, which can subsequently be used to make appropriate risk management decisions. Risk characterization is the risk assessment step in which the questions of the risk managers are directly addressed. The risk characterization can often include one or more estimates of risk, risk descriptions, and evaluations of risk management options. Those estimates may include economic and other evaluations in addition to estimates of risk attributable to the management options (FAO and WHO 2021). Although the Codex risk assessment framework is a common context for undertaking risk characterization, it is by no means the only context. In actual practice, an assessment of the risk may include some or all of these steps. The scientific analysis comprising any one of these steps may be sufficient on their own for decision-making. Risk assessments can follow a bottom-up or top-down approach. A bottom-up approach links knowledge about the prevalence and concentration of a hazard in a food source with knowledge about the causal pathways, transmission routes and dose–response relations. Alternatively, top-down approaches use observational epidemiological information to assess risk, typically making use of statistical regression models (FAO and WHO 2021). Examples of the risk assessment process according to FAO/WHO guideline can be found listed in the EFSA publication "Scientific Opinion on an Update on the Present Knowledge on the Occurrence and Control of Foodborne Viruses" (EFSA 2011a).

1.3 Considerations for risk assessment for viruses in food

The same principles of QMRA apply for all pathogens, but there are important reasons to treat viruses as a separate category of hazard (CAC 2012, EPA/USDA-FSIS 2012, FAO and WHO 2008b). Microbial risk assessment emerged as a separate discipline to chemical risk assessment by the mid-1990s (Bradshaw and Jaykus 2016, Foegeding et al. 1994, Nauta 2021). Since 2008, the FAO and WHO have formally acknowledged viruses in food as a hazard category of importance (FAO and WHO 2008b). Food safety measures for bacteria are insufficient in dealing with virus risk, due to some key difference in persistence and infectivity (Bradshaw and

Jaykus 2016, FAO and WHO 2008b). This can lead to diverging standards for virus food safety criteria. Some of the earliest examples of published MRAs focused on viruses as a hazard (Gerba et al. 1996b, Gerba and Haas 1988, Haas 1983, Haas et al. 1993). Since then, however, most published risk assessments have been for bacterial hazards. Virus hazards do not require a new paradigm of risk, just some differences in detail, similar to the separation between chemical and microbial hazards (Bradshaw and Jaykus 2016, Havelaar and Rutjes 2008). Rather, the problem is a lack of data on certain key question, and the difficulty in obtaining this data by present detection. The specific details for each stage of risk assessment will be presented in the sections that follow.

1.3.1 Hazard identification for viruses in food

The hazard identification step for virus risk assessment is a qualitative description of a food product, a virus, and the effect of both on a population of concern. Viruses are estimated to cause more than 50% of the world's foodborne disease burden (Havelaar et al. 2015, Koopmans and Duizer 2004). The most significant viruses, as identified by the FAO/WHO, are Hepatitis A, Norovirus, and rotavirus, with emerging viruses like Hepatitis E also of particular interest (FAO and WHO 2008b). The foods most associated with virus outbreaks are shellfish, fresh produce, and ready-to-eat or pre-prepared foods (Duizer and Koopmans 2008, Le Guyader and Atmar 2008). These were identified as the virus-food pairings of greatest concern by the FAO/WHO (FAO and WHO 2008b). Foods of animal origin are also of concern, particularly Hepatitis E virus in pigmeat (De Roda Husman and Bouwknegt 2013, Deest et al. 2007, Harrison and DiCaprio 2018, Matsubayashi et al. 2008, Treagus et al. 2021).

1.3.2 Exposure assessment for viruses in food

The exposure assessment stage for virus risk assessment has a specific set of concerns and challenges that are important to consider. Since most viruses of concern do not grow outside of a human or animal host, the main factors to consider will be transmission, detection, inactivation, survival or persistence and consumption. Much data for exposure modelling falls into these categories (Bradshaw and Jaykus 2016, FAO and WHO 2008b). Viruses come into contact with food, either through wastewater contamination, unhygienic handling, or zoonotically in pigmeat, and bind or attach to the food matrix (EFSA 2011a, Le Guyader and Atmar 2008). Viruses are shed in high quantities by the infected and have a low median infectious dose (Koopmans et al. 2003, Porsbo et al. 2013, Seymour and Appleton 2001). High survival rates and low infectious doses, compared with bacteria, mean that once food is contaminated an inactivation or removal step will be necessary to avoid outbreak events. The main data gaps in establishing the exposure assessment model are detecting infectious copies, and modelling inactivation and survival rates (De Roda Husman and Bouwknegt 2013, FAO and WHO 2008a, Jansen 2008). The effect of the food matrix, both in binding to the virus and in inhibiting infectivity, is also of concern. The goal is to provide an estimated number of units consumed, using quantitative risk assessment in combination with the tools of predictive microbiology.

1.3.3 Transmission of viruses in food

The transmission pathway for a foodborne virus is defined first by food coming into contact with a virus reservoir, then virus attaching to food, and finally persisting until ingestion (Le Guyader and Atmar 2008). The principal transmission routes identified by the FAO/WHO committee are wastewater contamination, handling by the infected, and animal or zoonotic sources (FAO/WHO 2008a). The main foods of concern for virus transmission are fresh produce, ready-to-eat food, shellfish, with pigmeat being an emerging concern for zoonotic spread. For fresh produce, virus is adsorbed onto the surface of fruits and vegetables and has also been observed getting taken directly into the plant tissue (Chancellor et al. 2006, Katzenelson and Mills 1984). During production and preparation of food, viruses can easily transfer between hands, surfaces, and foods. Depending on the hazard, data needed for modelling may be limited (De Roda Husman and Bouwknegt 2013). Shellfish accumulate viruses from the environment by filtering seawater that has been exposed to wastewater and concentrating copies in tissue via the digestive tract (McLeod et al. 2009, Schwab et al. 1998, Wang et al. 2008). The rate of shellfish accumulation can be affected by environmental factors as well as specific virus and shellfish (Burkhardt and Calci 2000, Le Guyader and Atmar 2008). The difference between virus concentrations at point of detection and point of consumption is negligible unless inactivation steps are included in the processing chain (Havelaar and Rutjes 2008). This suggests that risk management of virus hazards should focus both on modelling of inactivation and on the prevention of initial contact through better hygienic controls.

1.3.4 Detection of viruses in food

Detection of a foodborne hazard is critical to exposure assessment and a particular challenge for viruses. The increasing awareness of viruses as an important foodborne hazard is in part due to better methods for detecting and identifying them. As detection methods have improved, so have risk management outcomes. Le Guyader and Atmar (2008) give the example that in identifying the digestive glands as principal reservoir for shellfish accumulation, the remaining flesh can be discarded during monitoring, allowing for more accurate detection. Foodborne virus is detected in three steps (Havelaar and Rutjes 2008). First, the food matrix is treated to allow extraction or release of nucleic acids. This is followed by concentration of virus from the extract. Finally, the concentrated virus can be detected directly, either by molecular amplification (which can be quantitative or not) or by culturing in cell lines. The absence of reliable detection and quantification methods for the most important viruses is what limited virus risk assessment up until the FAO/WHO committee report of 2008. Presence or absence testing is insufficient for the stochastic distribution fitting needed for full (De Roda Husman and Bouwknegt 2013). Subsequent adoption of PCR based detection methods have allowed for some development of quantitative virus risk assessments, but the remaining limitations still lead to significant data gaps. The most critical hazards, norovirus and hepatitis viruses, have not allowed for reliable cell culturing from food matrices (Bosch et al. 2018, De Medici et al. 2001, Di Cola et al. 2021, Duizer 2004, Ettayebi et al. 2021, Ludwig-Begall et al. 2021, Randazzo and Sánchez 2020). This means there is little to no data on the true level of infectious virus, as the molecular methods do not distinguish between infectious genome copies and non-infectious (Richards 1999, Zhang et al. 2019). The ratio between non-infectious and infectious copies is variable and uncertain, especially over longer periods of time (Gassilloud and Gantzer 2005). This can vary from being almost entirely infectious to having a 50,000:1 ratio of non-infectious to infectious (Duizer et al. 2004, Gassilloud and Gantzer 2005, Havelaar and Rutjes 2008, Rutjes et al. 2005). Clearly this is a significant issue for accurate risk assessment, and likely to lead to overestimation of risk if detection results are taken at face value. More sophisticated detection methods have been developed in recent years that overcome these limitations but are not yet in common use. The most common source of data to overcome the lack of data on infectivity are the use of surrogate viruses, an approach which comes with its own downsides and limitations (Atmar et al. 2008, CAC 2012, FAO and WHO 2008b, HPA 2004, Richards 2012).

1.3.5 Surrogates for viruses in food

The use of data from surrogate viruses as a proxy for virus hazards of interest has to date been a necessary part of virus risk assessment. Surrogates differ from indicator organisms used for risk control, like the use of *E. coli* as an indicator for faecal pollution in shellfish in EU regulations (as established, insufficient for virus risk control) (Campos and Lees 2014, Doré et al. 2000, Potasman et al. 2002). Surrogates are viruses that can be cultured and therefore used as a substitute source of data for those that cannot. They allow assumptions to be made about the infectivity of the target virus when detected by molecular means, and about the effectiveness of inactivation methods. The surrogate virus is not itself relevant from the perspective of risk management but is assumed to behave similarly to the real virus of interest, like norovirus. Surrogates are chosen based on their similarity to the target virus, both in molecular structure and, where relevant, environmental presence (CAC 2012). The main limitation of surrogate viruses is uncertainty over the true similarity between the surrogate and the target, which cannot be directly tested. Viruses can react differently to inactivation methods like disinfection, and the effect of the food matrix can have differing impacts on different viruses, even between species of host like shellfish.

1.3.6 Persistence of viruses in food

Persistence over time is much more relevant for virus exposure assessment than bacteria, for several reasons (Bradshaw and Jaykus 2016, FAO and WHO 2008b). Virus copies will not replicate or grow in food, meaning that the survival rate is the only thing affecting the final concentration at consumption. In addition, viruses are infectious at lower doses and are likely to contaminate a large numbers of shed copies at point of contact (Atmar et al. 2008, Teunis et al. 2008). The absence of metabolism also makes viruses more persistent than bacteria. Shellfish have been observed to retain infectious norovirus for several weeks after accumulation, much longer than similar observations in case of bacteria (Campos and Lees 2014, Choi and Kingsley 2016). Other notable virus persistence rates observed in food include Hepatitis A in fresh produce (Croci et al. 2002, Sun et al. 2012). Low storage temperatures, intended to prevent bacterial growth, will help to preserve virus integrity during the food processing chain (Bidawid et al. 2001, Hernroth and Allard 2007, Kingsley and Richards 2003, Le Guyader and Atmar 2008, Ward and Irving 1987). The food matrix is thought to affect viral persistence, given the role of binding and attachment, though there is little concrete data on how, owing to detection difficulties (Le Guyader and Atmar 2008). Given this persistence, the likelihood of exposure at consumption if food is contaminated at any point in the chain, is high. Inactivation or removal during processing is therefore necessary to avoid outbreaks for foods like produce and shellfish, which are consumed raw or lightly cooked. The relevance for QMRA overall is in modelling the post-harvest stages and the potential for cross-contamination.

1.3.7 Inactivation of viruses in food

If, during the farm-to-fork process, food acquires viruses in sufficient numbers to cause infection, some form of inactivation treatment will be necessary to avoid outbreak events. Without an inactivation stage, the high persistence of viruses mean there will be negligible difference between concentrations early in the food processing chain and concentrations at consumption (Havelaar and Rutjes 2008). Inactivation is any treatment that aims to reduce infectious virus in a food, either by destroying the capacity to infect or by removing the virus entirely. As discussed in an earlier chapter, inactivation methods can be thermal, based on heat treatment or cooking, or non-thermal, based on various other physical, chemical or biological methods. Removal for produce is based on washing in water or disinfectant. In shellfish, as the virus accumulates internally, removal is more challenging. Shellfish must be placed in uncontaminated waters, either at a clean farm location, or in an artificially sterilized tank environment. They can then over time purge contamination through the digestive process. The effectiveness of these methods is unpredictable and unreliable (Le Guyader and Atmar 2008, McLeod et al. 2017, Polo et al. 2014). Difficulties in detection make the efficacy hard to judge, and outbreaks are still observed following treatments. Deactivation treatments exist that are not accurately assessed by molecular methods, like forms of cooking and UV exposure (Duizer et al. 2004, Zhang et al. 2019). The primary challenge remains the determination of the ratio of infectious to non-infectious virus, which is ultimately an issue with detection.

1.3.8 Consumption of viruses in food

The assumptions made when modelling consumption can have a big impact on virus risk assessment, due to the general nature of the hazards under consideration. Foods most associated with virus outbreaks are often lightly cooked or consumed raw: fresh produce, shellfish, ready to eat foods, and cured meats (Havelaar and Rutjes 2008). Consumption assumptions do not otherwise differ from standard MRA practices. For less commonly consumed foods like shellfish, survey data may be unreliable for estimating consumption for individuals, and a per person serving risk estimate may be more appropriate to begin with (FAO and WHO 2008a, Havelaar and Rutjes 2008). Consumption data can be found through consumer surveys, and located in databases like the EFSA Comprehensive European Food Consumption Database (EFSA 2011b). Important data for quantitative modelling on the frequency of specific food preparation methods can be scarce. This is typical in risk assessment and not specific to viral hazards (De Roda Husman and Bouwknegt 2013).

1.4 Hazard characterization for viruses in food

1.4.1 Mechanism of infection

Hazard characterizations for virus hazards will share a set of important assumptions. A dose-response model is used to quantitatively characterize the effect of hazard on a person who is once exposed. This model gives the mathematical relationship between the quantity of infectious units consumed and the probability of infection or illness (CAC 1999, Haas et al. 2014). For viruses, the assumption behind the most dose-response models is that a single virus copy is sufficient to initiate infection, if it can navigate the host's immune system. Additionally, viruses are assumed to act independently, neither hindering nor helping other virus copies along the way. Higher virus loads do not weaken or overwhelm the immune system in this assumption, except in allowing more chances to cause infection (FAO and WHO 2003, Haas et al. 2014). A limited number of models are generally used to cover these "single hit" assumptions. The most common are the exponential, the hypergeometric and the Beta-Poisson (Haas 1983, Teunis and Havelaar 2000, Zwietering and Havelaar 2006). These models include assumptions about both the distribution of the pathogen and the host immune response and should be harmonized with the output of the exposure assessment in a full risk characterization. Haas originally modelled the probability of infection following exposure to a single virus copy as a Beta (Haas 1983), which can cover both host-to-host and virus-to-virus variability (Nauta 2021, Pouillot et al. 2015). This assumption has been held for the majority of published virus hazard characterization or substituted with a fixed probability. Dose response models are fit for a hazard based on the data available, whether this is from challenge studies, outbreak data, animal trials, or surrogates (Nauta 2021). The impact of the food matrix is important to consider, and there can be limited data available on this for a specific hazard (Havelaar and Rutjes 2008, Nauta 2021).

1.4.2 Data availability

The dose-response relationship for viral hazards is generally more difficult to assess than that of bacteria. Data may be limited and difficult to interpret where available, due to the uncertainty on whether the measured doses are infectious or not. Few enteric viruses have data available for dose-response modelling, and where it is available, the data is still challenging to interpret. Major viruses with challenge studies available include norovirus, hepatitis A virus, rotavirus, poliovirus and echovirus-12, a common surrogate for hepatitis (Bradshaw and Jaykus 2016). Rotavirus and norovirus both have low median infectious doses estimated in the literature, 10 and 13 copies respectively (Teunis et al. 2008, Ward et al. 1986). Dose-response models based directly on shellfish borne norovirus outbreaks have shown similar results, and the outbreak data available generally supports high infectivity at low levels of virus (Thebault et al.

2013). The difficulties in distinguishing infectious virus copies from non-infectious ones also make interpreting challenge studies difficult, given the uncertainty over the true dose. Some challenge studies suggest aggregation of virus particles may also reduce infectivity, though in practice this is impossible to distinguish from the effect of non-infectious particles (Atmar et al. 2015, Kirby et al. 2015, Teunis et al. 2008). The food matrix for a specific hazard will also have some unknown impact on the model (De Roda Husman and Bouwknegt 2013). Additionally, challenge studies are performed on healthy volunteers, which may not reflect the subpopulations of interest (Gerba et al. 1996a). All this considered however, the uncertainties of the virus dose-response model are still not as significant as those of exposure assessment, given the high variance in infectivity.

1.5 Comparison between viral and bacterial QMRA in oysters

The number of virus risk assessments published to date is outnumbered by bacterial risk assessments due to the easier availability of data over time. As discussed earlier, some of the earliest MRAs published were on virus hazards (Gerba et al. 1996b, Gerba and Haas 1988, Haas 1983, Rose and Sobsey 1993). In the last decade, more and more specific virus risk assessments have been published, covering a wider variety of foodborne hazards. Comparing and contrasting a virus MRA and a bacterial MRA for the same food product will demonstrate some of the key differences between the two.

Shellfish are associated with several microbial hazards and offer a good opportunity to compare the needs of bacterial QMRA with those of virus QMRA. Because oysters are mainly consumed raw, they pose a higher risk of causing illness to consumers and are therefore a product of interest to risk assessors. Two published risk assessments showcase the main points of difference between the two categories of hazard. The bacterial risk assessment, prepared by the FAO and WHO was first published in 2005 and examines the risk of *Vibrio vulnificus* in oysters (FAO and WHO 2011). This work was part of a package of risk assessments, looking at the impact on public health of *Vibrio* spp. in raw molluscan shellfish in general, based on available data. The ultimate objective was to evaluate ways to mitigate the risk of *Vibrio* related septicemia caused by raw oyster consumption. The viral risk assessment comes almost two decades later, in 2021. It looks at norovirus illness caused by raw oyster consumption and was prepared jointly by the US and Canadian food safety authorities (Pouillot et al. 2021). Here the focus was on modelling and comparing the impact of environmental factors, rather than comparing mitigation methods directly. Both risk assessments follow the standard Codex paradigm, answering many of the same questions in each.

Comparing the two, several interesting differences emerge for each point in the risk assessment process:

- In the hazard identification, *V. vulnificus* is identified as a bacteria indigenous to warmer estuarine environments, while norovirus is a human-specific virus introduced to estuaries through the wastewater system. *V. vulnificus* has a much lower rate of infection, mainly affecting immunocompromised individuals, with a high case-fatality rate of 50%. The risk assessment for the bacteria focuses on the specific clinical outcome of septicemia, aiming to mitigate this risk. For norovirus, there is no target population, as the biggest impact to public health comes from the much higher rates of infection.

- The exposure assessments for both hazards are similar in their framework, linking a pre-harvest accumulation phase, the point of harvest, and the point of consumption. For both risk assessments, the most important outcome is the prevalence and concentration at the point of harvest. There is no potential for cross-contamination considered either. However, given that norovirus will not grow during post-harvest storage, and has high persistence, little attention is paid to the post-harvest portion of the framework. This contrasts with the *V. vulnificus* assessment, where post-harvest growth and storage conditions take up most of the modelling parameters.

- In general, the influence of temperature is inverted. Like many bacteria, *V. vulnificus* thrives in warmer conditions, both in the environment and during post-harvest storage. This causes a strong seasonal effect, with risks highest in warmer months. For norovirus, the effect of temperature is more complex, with higher temperatures leading to higher rates of accumulation in the oyster but also higher rates of elimination. The ultimate seasonal effect on norovirus is the opposite to *Vibrio*, with winter months posing more risk to consumers.

- In both exposure assessments, the assumptions made for consumption are identical, with the same consumption data used for modelling. For less common food like oysters, there is limited consumption data, especially when considering target subpopulations.

- The difference in dose-response modelling is the consideration of the target population; with the *V. vulnificus* risk assessment aiming to reduce fatalities in immunocompromised individuals. This is estimated below 7% of the total consumer population. For norovirus, the impact is felt in the number of illnesses, rather than the impact of individual fatalities. The higher infectivity of norovirus (median infectious dose several orders of magnitude lower than for *V. vulnificus*) makes prevalence rather than concentration as the main concern.

- Susceptibility to norovirus is based largely on genetic profile, with the susceptible proportion of the population (~ 75%) highly likely to be infected once exposed to infectious virus. For *V. vulnificus*, susceptibility depends on the host's overall immune strength. In both cases, this susceptibility will vary depending on the demographics of the region in question, but for different reasons.

- The *V. vulnificus* risk characterization concludes with a worst-case estimate of under 40 cases of septicemia for the gulf region of the United States. The number of norovirus cases annually is estimated in the QMRA to be over 70,000 across the US and Canada, although there is a high uncertainty due to the unknown impact of infectivity.

- The *V. vulnificus* risk assessment concludes that the assumptions and framework that it uses can be re-used for similar bacterial hazards. There is no similar conclusion for the application of the norovirus model to other enteric virus hazards, although a correlation is identified between a surrogate virus MSC and the risk of norovirus illness.

The key differences between QMRA for bacteria and for viruses discussed in this chapter can be seen in these examples. The high infectivity of viruses and the differences in persistence and inactivation show the different considerations needed compared with most bacterial hazards.

1.6 Future trends in virus risk assessment

1.6.1 Major lacks

To date, the main thing preventing greater availability of virus QMRAs is a simple lack of data. The inability to cultivate the most significant viruses has meant little to no direct data on the prevalence and concentration of infectious viruses at all stages in the food production chain. It has also meant a lack of data on the efficacy of inactivation and removal methods and the transfer rates between handlers, surfaces and foods. To a lesser extent of importance, it has also made it difficult to fit accurate dose-response models. Molecular methods are likelier to suffer from limits of detection than culturing methods, which is significant for viruses like norovirus where doses below those limits can cause infection (Rutjes et al. 2006). There is little data on variability between host-to-host susceptibility and virus-to-virus virulence, outside of crude challenge studies.

1.6.2 Major challenges

There are therefore challenges to overcome before QVRA data can have the same standard as bacterial ones. Chiefly better detection methods, and better sensitivity to infectious viruses. This will lead to better information on the true concentrations likely in contaminated food and the true effect of inactivation or removal methods like shellfish depuration and produce disinfection. In all cases where better data is needed—prevalence, transmission, transfer, persistence, inactivation—better detection methods will contribute to a solution (Bradshaw and Jaykus 2016). Even with new methods of detection and culturing, surrogates will still have a place in experimental and monitoring work. The relationship between surrogate viruses and the hazards they represent will however also be better illuminated with better data and better detection methods. More information on virus and host variation in hazards will lead to more sophisticated and more accurate risk assessment models.

1.6.3 Recent developments

Some promising developments have appeared in recent years, which may serve to meet the challenges ahead. Cell culturing methods for hepatitis A and hepatitis E virus have been integrated with molecular methods for assessing infectivity from food samples (Bosch et al. 2018, Cook et al. 2017, Randazzo and Sánchez 2020). An *in-vitro* cultivation for norovirus infectivity based on human intestinal enteroids was published in 2016, and continues to be developed for food and environmental sampling (Estes et al. 2019, Ettayebi et al. 2021, 2016, Overbey et al. 2021). PCR-based quantification has also become more sensitive, with viability PCR techniques using pre-treatment with viability markers (enzymes or dyes) to assess capsid integrity and better represent infectivity levels (Chen et al. 2020, Di Cola et al. 2021, Elizaquível et al. 2014, Moreno et al. 2015). Tools like whole genome sequencing and metagenomics offer new insights in hazard identification (Collineau et al. 2019, EFSA BIOHAZ et al. 2019, Franz et al. 2016, Fritsch et al. 2018, Rantsiou et al. 2018), hazard characterization (Haddad et al. 2018) and exposure assessment (den Besten et al. 2018). Similarly, tools for risk assessment have improved, with better data sharing, better software resources, predictive modelling databases and harmonization of concepts across risk assessment (Bassett et al. 2012, Plaza-Rodríguez et al. 2018). The FAO/WHO Joint Expert Meetings on Microbiological Risk Assessment (JEMRA) continue to publish and coordinate the best expert advice on the topic, as do bodies like EFSA and the USDA (LeJeune et al. 2021). It will take time for these methods to propagate widely. The time and costs needed to use them may render them unsuitable for routine monitoring work (EFSA BIOHAZ et al. 2019, Nauta 2021). However, it is evident that there is now the potential for a new generation of risk assessment, where viruses will play a bigger role.

2. Concluding remarks

The problem of virus risk assessment has been identified for decades, and only until recently the opportunity to meet the challenge has begun. Given the contribution of viruses to the foodborne disease burden, greater attention should be given to them in the coming years. The output rate of virus risk assessment will continue to increase as more data becomes available. The control of virus hazards will improve by greater awareness of virus attachment and virus inactivation, especially with the possible variability. Microbiological control criteria are always needed, and with the next generation of tools and data, the future of QMRA could correctly address the role of enteric viruses as significant foodborne hazards.

Establishing systems of control and management, specific to viral hazards is both important and challenging. Standardized guidelines on control criteria for managing virus risk are important for international food trade (Jansen 2008). In response to these requirements, risk assessments for specific hazards are needed. Existing bacterial criteria are not sufficient to control virus risk. Viruses are more persistent than bacteria on food, are infectious in lower numbers, and the efficacy of virus inactivation or removal methods is uncertain. In shellfish, microbiological criteria are established in Codex guidelines, based on *E. coli* concentrations as an indicator organism (Campos and Lees 2014, Doré et al. 2000, Potasman et al. 2002). This is poorly correlated with virus risk and has resulted in inconsistent virus safety standards being applied by different countries (Jansen 2008). The absence of knowledge on infectivity and dose-response means that risk analysis uses risk assessment results as a "worst case scenario" and relies on scenario analysis and relative risk estimation for risk management. This allows sensitivity analysis and guides future data but overestimates risk in aggregate (Havelaar and Rutjes 2008).

References

Atmar, R.L., Opekun, A.R., Gilger, M.A., Estes, M.K., Crawford, S.E., Neill, F.H. et al. 2008. Norwalk virus shedding after experimental human infection. Emerging Infect. Dis. 14: 1553–1557. https://doi.org/10.3201/eid1410.080117.

Atmar, R.L., Opekun, A.R., Estes, M.K. and Graham, D.Y. 2015. Reply to Kirby et al. J. Infect. Dis. 211: 167–167. https://doi.org/10.1093/infdis/jiu382.

Bassett, John, Nauta, M., Lindqvist, R. and Zwietering, M. 2012. Tools for Microbiological Risk Assessment (Report), Tools for Microbiological Risk Assessment. ILSI Europe.

Bidawid, S., Farber, J.M. and Sattar, S.A. 2001. Survival of hepatitis A virus on modified atmosphere-packaged (MAP) lettuce. Food Microbiology 18: 95–102. https://doi.org/10.1006/fmic.2000.0380.

Bosch, A., Gkogka, E., Le Guyader, F.S., Loisy-Hamon, F., Lee, A., van Lieshout, L. et al. 2018. Foodborne viruses: Detection, risk assessment, and control options in food processing. International Journal of Food Microbiology 285: 110–128. https://doi.org/10.1016/j.ijfoodmicro.2018.06.001.

Bradshaw, E. and Jaykus, L.-A. 2016. Risk assessment for foodborne viruses. pp. 471–503. *In*: Goyal, S.M. and Cannon, J.L. (eds.). Viruses in Foods, Food Microbiology and Food Safety. Springer International Publishing. https://doi.org/10.1007/978-3-319-30723-7_17.

Burkhardt, W. and Calci, K.R. 2000. Selective accumulation may account for shellfish-associated viral illness. Applied and Environmental Microbiology 66: 1375–1378. https://doi.org/10.1128/AEM.66.4.1375-1378.2000.

CAC. 1999. Principles and guidelines for the conduct of microbiological risk assessment CAC/GL 30-1999 7.

CAC. 2012. Guidelines on the application of general principles of food hygiene to the control of viruses in food. CAC/GL 79-2012. Codex Alimentarius Commission.

Campos, C.J.A. and Lees, D.N. 2014. Environmental transmission of human noroviruses in shellfish waters. Appl. Environ. Microbiol. 80: 3552–3561. https://doi.org/10.1128/AEM.04188-13.

Chancellor, D.D., Tyagi, S., Bazaco, M.C., Bacvinskas, S., Chancellor, M.B., Dato, V.M. et al. 2006. Green onions: potential mechanism for hepatitis a contamination. Journal of Food Protection 69: 1468–1472. https://doi.org/10.4315/0362-028X-69.6.1468.

Chen, J., Wu, X., Sánchez, G. and Randazzo, W. 2020. Viability RT-qPCR to detect potentially infectious enteric viruses on heat-processed berries. Food Control 107: 106818. https://doi.org/10.1016/j.foodcont.2019.106818.

Choi, C. and Kingsley, D.H. 2016. Temperature-dependent persistence of human norovirus within oysters (Crassostrea virginica). Food Environ. Virol. 8: 141–147. https://doi.org/10.1007/s12560-016-9234-8.

Collineau, L., Boerlin, P., Carson, C.A., Chapman, B., Fazil, A., Hetman, B. et al. 2019. Integrating whole-genome sequencing data into quantitative risk assessment of foodborne antimicrobial resistance: a review of opportunities and challenges. Frontiers in Microbiology 10.

Cook, N., D'Agostino, M. and Johne, R. 2017. Potential approaches to assess the infectivity of hepatitis E virus in pork products: a review. Food Environ. Virol. 9: 243–255. https://doi.org/10.1007/s12560-017-9303-7.

Covello, V.T. and Mumpower, J. 1985. Risk analysis and risk management: an historical perspective. Risk Analysis 5: 103–120. https://doi.org/10.1111/j.1539-6924.1985.tb00159.x.

Croci, L., De Medici, D., Scalfaro, C., Fiore, A. and Toti, L. 2002. The survival of hepatitis A virus in fresh produce. International Journal of Food Microbiology 73: 29–34. https://doi.org/10.1016/S0168-1605(01)00689-4.

De Medici, D., Croci, L., Di Pasquale, S., Fiore, A. and Toti, L. 2001. Detecting the presence of infectious hepatitis A virus in molluscs positive to RT-nested-PCR. Letters in Applied Microbiology 33: 362–366. https://doi.org/10.1046/j.1472-765X.2001.01018.x.

De Roda Husman, A.M. and Bouwknegt, M. 2013. 8 - Quantitative risk assessment for food- and waterborne viruses. pp. 159–175. *In*: Cook, N. (ed.). Viruses in Food and Water, Woodhead Publishing Series in Food Science, Technology and Nutrition. Woodhead Publishing. https://doi.org/10.1533/9780857098870.2.159.

Deest, G., Zehner, L., Nicand, E., Gaudy-Graffin, C., Goudeau, A. and Bacq, Y. 2007. [Autochthonous hepatitis E in France and consumption of raw pig meat]. Gastroenterol. Clin. Biol. 31: 1095–1097. https://doi.org/10.1016/s0399-8320(07)78342-2.

den Besten, H.M.W., Amézquita, A., Bover-Cid, S., Dagnas, S., Ellouze, M., Guillou, S., Nychas, G., O'Mahony, C., Pérez-Rodriguez, F. and Membré, J.-M. 2018. Next generation of microbiological risk assessment: Potential of omics data for exposure assessment. International Journal of Food Microbiology, Omics in MRA - The Integration of Omics in Microbiological Risk Assessment 287: 18–27. https://doi.org/10.1016/j.ijfoodmicro.2017.10.006.

Di Cola, G., Fantilli, A.C., Pisano, M.B. and Ré, V.E. 2021. Foodborne transmission of hepatitis A and hepatitis E viruses: A literature review. International Journal of Food Microbiology 338: 108986. https://doi.org/10.1016/j.ijfoodmicro.2020.108986.

Doré, W.J., Henshilwood, K. and Lees, D.N. 2000. Evaluation of F-specific RNA bacteriophage as a candidate human enteric virus indicator for bivalve molluscan shellfish. Applied and Environmental Microbiology 66: 1280–1285. https://doi.org/10.1128/AEM.66.4.1280-1285.2000.

Duizer, E. 2004. Laboratory efforts to cultivate noroviruses. Journal of General Virology 85: 79–87. https://doi.org/10.1099/vir.0.19478-0.

Duizer, E., Bijkerk, P., Rockx, B., de Groot, A., Twisk, F. and Koopmans, M. 2004. Inactivation of caliciviruses. Applied and Environmental Microbiology 70: 4538–4543. https://doi.org/10.1128/AEM.70.8.4538-4543.2004.

Duizer, E. and Koopmans, M. 2008. Emerging food-borne viral diseases. Food-borne Viruss: Progress and Challenges 117–145. https://doi.org/10.1128/9781555815738.ch5.

EFSA. 2011a. Scientific Opinion on an update on the present knowledge on the occurrence and control of foodborne viruses. EFSA Journal 9, n/a-n/a. https://doi.org/10.2903/j.efsa.2011.2190.

EFSA. 2011b. Use of the EFSA Comprehensive European Food Consumption Database in Exposure Assessment. EFSA Journal 9, 2097. https://doi.org/10.2903/j.efsa.2011.2097.

EFSA BIOHAZ, E.P. on B., Koutsoumanis, K., Allende, A., Alvarez-Ordóñez, A., Bolton, D., Bover-Cid, S., Chemaly, M. et al. 2019. Whole genome sequencing and metagenomics for outbreak investigation, source attribution and risk assessment of food-borne microorganisms. EFSA Journal 17: e05898. https://doi.org/10.2903/j.efsa.2019.5898.

Elizaquível, P., Aznar, R. and Sánchez, G. 2014. Recent developments in the use of viability dyes and quantitative PCR in the food microbiology field. J. Appl. Microbiol. 116: 1–13. https://doi.org/10.1111/jam.12365.

EPA/USDA-FSIS. 2012. Microbial risk assessment guidelines: pathogenic microorganisms with focus on food and water, EPA/100/J-12/001; USDA/FSIS/2012-001. Prepared by the Interagency Microbiological Risk Assessment Guideline Workgroup.

Estes, M.K., Ettayebi, K., Tenge, V.R., Murakami, K., Karandikar, U., Lin, S.-C. et al. 2019. Human norovirus cultivation in nontransformed stem cell-derived human intestinal enteroid cultures: success and challenges. Viruses 11: 638. https://doi.org/10.3390/v11070638.

Ettayebi, K., Crawford, S.E., Murakami, K., Broughman, J.R., Karandikar, U., Tenge, V.R. et al. 2016. Replication of human noroviruses in stem cell–derived human enteroids. Science 353: 1387–1393. https://doi.org/10.1126/science.aaf5211.

Ettayebi, K., Tenge, V.R., Cortes-Penfield, N.W., Crawford, S.E., Neill, F.H., Zeng, X.-L., Yu, X., Ayyar, B.V., Burrin, D., Ramani, S., Atmar, R.L. and Estes, M.K. 2021. New insights and enhanced human norovirus cultivation in human intestinal enteroids. mSphere 6: e01136–20. https://doi.org/10.1128/msphere.01136-20.

FAO, WHO. 1995. Application of risk analysis to food standards issues : report of the Joint FAO/WHO expert consultation, Geneva, Switzerland, 13–17 March 1995 (No. WHO/FNU/FOS/95.3. Unpublished). World Health Organization.

FAO, WHO. 2003. Hazard Characterization for Pathogens in Food and Water: Guidelines. Food & Agriculture Org.

FAO, WHO. 2008a. Exposure Assessment of microbiological Hazards in Food.

FAO, WHO. 2008b. Viruses in Food: Scientific Advice to Support Risk Management Activities: Meeting Report. Microbiological Risk Assessment Series (MRA) 13, Microbiological Risk Assessment Series (FAO/WHO). FAO/WHO, Rome, Italy.

FAO, WHO. 2009. Risk Characterization of Microbiological Hazards in Food.

FAO, WHO. 2011. Risk Assessment of Vibrio Vulnificus in Raw Oysters: Interpretative Summary and Technical Report. World Health Organization.

FAO, WHO. 2019. Codex Alimentarius Commission – Procedural Manual. 27 ed., Codex Alimentarius - Joint FAO/WHO Food Standards Programme. FAO, Rome, Italy.

FAO and WHO. 2021. Microbiological Risk Assessment—Guidance for food, Microbiological Risk Assessment Series. FAO and WHO, Rome, Italy. https://doi.org/10.4060/cb5006en.

Foegeding, P.M., Roberts, T., Bennett, J., Bryan, F.L., Cliver, D.O., Doyle, M.P. et al. 1994. Foodborne pathogens: risks and consequences. Task force report (Council for Agricultural Science and Technology) (USA).

Franz, E., Gras, L.M. and Dallman, T. 2016. Significance of whole genome sequencing for surveillance, source attribution and microbial risk assessment of foodborne pathogens. Current Opinion in Food Science, Food Microbiology • Functional Foods and Nutrition 8: 74–79. https://doi.org/10.1016/j.cofs.2016.04.004.

Fritsch, L., Guillier, L. and Augustin, J.-C. 2018. Next generation quantitative microbiological risk assessment: Refinement of the cold smoked salmon-related listeriosis risk model by integrating genomic data. Microbial Risk Analysis, Special issue on 10th International Conference on Predictive Modelling in Food: Interdisciplinary Approaches and Decision-Making Tools in Microbial Risk Analysis 10: 20–27. https://doi.org/10.1016/j.mran.2018.06.003.

Gassilloud, B. and Gantzer, C. 2005. Adhesion-aggregation and inactivation of poliovirus 1 in groundwater stored in a hydrophobic container. Applied and Environmental Microbiology 71: 912–920. https://doi.org/10.1128/AEM.71.2.912-920.2005.

Gerba, C.P., Rose, J.B. and Haas, C.N. 1996a. Sensitive populations: who is at the greatest risk? International Journal of Food Microbiology, Risk Analysis and Production of Safe Food 30: 113–123. https://doi.org/10.1016/0168-1605(96)00996-8.

Gerba, C.P., Rose, J.B., Haas, C.N. and Crabtree, K.D. 1996b. Waterborne rotavirus: A risk assessment. Water Research 30: 2929–2940. https://doi.org/10.1016/S0043-1354(96)00187-X.

Gerba, C.P. and Haas, C.N. 1988. Assessment of risks associated with enteric viruses in contaminated drinking water. pp. 489–494. *In*: Lichtenberg, J.J. (ed.). Chemical and Biological Characterization of Municipal Sludges, Sediments, Dredge Spoils, and Drilling Muds. ASTM International.

Haas, C.N. 1983. Estimation of risk due to low doses of microorganisms: a comparison of alternative methodologies. Am. J. Epidemiol. 118: 573–582.

Haas, C.N., Rose, J.B., Gerba, C. and Regli, S. 1993. Risk assessment of virus in drinking water. Risk Analysis 13: 545–552. https://doi.org/10.1111/j.1539-6924.1993.tb00013.x.

Haas, C.N., Rose, J.B. and Gerba, C.P. 2014. Quantitative Microbial Risk Assessment. John Wiley & Sons.

Haddad, N., Johnson, N., Kathariou, S., Métris, A., Phister, T., Pielaat, A. et al. 2018. Next generation microbiological risk assessment—Potential of omics data for hazard characterisation. International Journal of Food Microbiology, Omics in MRA - the Integration of Omics in Microbiological Risk Assessment 287: 28–39. https://doi.org/10.1016/j.ijfoodmicro.2018.04.015.

Harrison, L. and DiCaprio, E. 2018. Hepatitis E virus: an emerging foodborne pathogen. Frontiers in Sustainable Food Systems 2.

Havelaar, A.H. and Rutjes, S.A. 2008. Risk Assessment of Viruses in Food: Opportunities and Challenges 221–236. https://doi.org/10.1128/9781555815738.ch10.

Havelaar, A.H., Kirk, M.D., Torgerson, P.R., Gibb, H.J., Hald, T., Lake, R.J. et al., Devleesschauwer, B., Group, on behalf of W.H.O.F.D.B.E.R. 2015. World Health Organization global estimates and regional comparisons of the burden of foodborne disease in 2010. PLOS Med. 12: e1001923. https://doi.org/10.1371/journal.pmed.1001923.

Hernroth, B. and Allard, A. 2007. The persistence of infectious adenovirus (type 35) in mussels (*Mytilus edulis*) and oysters (*Ostrea edulis*). International Journal of Food Microbiology 113: 296–302. https://doi.org/10.1016/j.ijfoodmicro.2006.08.009.

HPA. 2004. Microbiological risk assessment for norovirus infection—contribution to the overall burden afforded by foodborne infections (No. 184-1-318). Porton Down.

Jansen, J. 2008. Use of the codex risk analysis framework to reduce risks associated with viruses in food. pp. 209–220. *In*: Food-Borne Viruses. John Wiley & Sons, Ltd, https://doi.org/10.1128/9781555815738.ch9.

Katzenelson, E. and Mills, D. 1984. Contamination of vegetables with animal viruses via the roots. Enteric Viruses in Water 15: 216–220. https://doi.org/10.1159/000409140.

Kingsley, D.H. and Richards, G.P. 2003. Persistence of hepatitis A virus in oysters. Journal of Food Protection 66: 331–334.

Kirby, A.E., Teunis, P.F. and Moe, C.L. 2015. Two human challenge studies confirm high infectivity of norwalk virus. J. Infect. Dis. 211: 166–167. https://doi.org/10.1093/infdis/jiu385.

Koopmans, M., Vennema, H., Heersma, H., van Strien, E., van Duynhoven, Y., Brown, D. et al. 2003. Early identification of common-source foodborne virus outbreaks in Europe. Emerg. Infect. Dis. 9: 1136–1142. https://doi.org/10.3201/eid0909.020766.

Koopmans, M. and Duizer, E. 2004. Foodborne viruses: an emerging problem. Int. J. Food Microbiol. 90: 23–41. https://doi.org/10.1016/s0168-1605(03)00169-7.

Lammerding, A.M. 1997. An overview of microbial food safety risk assessment†. Journal of Food Protection 60: 1420–1425. https://doi.org/10.4315/0362-028X-60.11.1420.

Le Guyader, F.S. and Atmar, R.L. 2008. Binding and inactivation of viruses on and in food, with a focus on the role of the matrix. pp. 189–208. *In*: Food-Borne Viruses. American Society of Microbiology.

LeJeune, J.T., Zhou, K., Kopko, C. and Igarashi, H. 2021. FAO/WHO Joint Expert Meeting on Microbiological Risk Assessment (JEMRA): Twenty Years of International Microbiological Risk Assessment. Foods 10: 1873. https://doi.org/10.3390/foods10081873.

Ludwig-Begall, L.F., Mauroy, A. and Thiry, E. 2021. Noroviruses—The state of the art, nearly fifty years after their initial discovery. Viruses 13: 1541. https://doi.org/10.3390/v13081541.

Matsubayashi, K., Kang, J.-H., Sakata, H., Takahashi, K., Shindo, M., Kato, M. et al. 2008. A case of transfusion-transmitted hepatitis E caused by blood from a donor infected with hepatitis E virus via zoonotic food-borne route. Transfusion 48: 1368–1375. https://doi.org/10.1111/j.1537-2995.2008.01722.x.

McLeod, C., Hay, B., Grant, C., Greening, G. and Day, D. 2009. Localization of norovirus and poliovirus in Pacific oysters. J. Appl. Microbiol. 106: 1220–1230. https://doi.org/10.1111/j.1365-2672.2008.04091.x.

McLeod, C., Polo, D., Saux, J.-C.L. and Guyader, F.S.L. 2017. Depuration and relaying: a review on potential removal of norovirus from oysters. Comprehensive Reviews in Food Science and Food Safety 16: 692–706. https://doi.org/10.1111/1541-4337.12271.

Moreno, L., Aznar, R. and Sánchez, G. 2015. Application of viability PCR to discriminate the infectivity of hepatitis A virus in food samples. International Journal of Food Microbiology 201: 1–6. https://doi.org/10.1016/j.ijfoodmicro.2015.02.012.

Nauta, M. 2021. Chapter 2 - Microbial food safety risk assessment. pp. 19–34. *In*: Morris, J.G. and Vugia, D.J. (eds.). Foodborne Infections and Intoxications (Fifth Edition). Academic Press. https://doi. org/10.1016/B978-0-12-819519-2.00015-3.

Overbey, K.N., Zachos, N.C., Coulter, C. and Schwab, K.J. 2021. Optimizing human intestinal enteroids for environmental monitoring of human norovirus. Food Environ. Virol. 13: 470–484. https://doi. org/10.1007/s12560-021-09486-w.

Plaza-Rodríguez, C., Haberbeck, L.U., Desvignes, V., Dalgaard, P., Sanaa, M., Nauta, M., Filter, M. and Guillier, L. 2018. Towards transparent and consistent exchange of knowledge for improved microbiological food safety. Current Opinion in Food Science, Food Chemistry and Biochemistry * Food Bioprocessing 19: 129–137. https://doi.org/10.1016/j.cofs.2017.12.002.

Polo, D., Álvarez, C., Díez, J., Darriba, S., Longa, Á. and Romalde, J.L. 2014. Viral elimination during commercial depuration of shellfish. Food Control 43: 206–212. https://doi.org/10.1016/j. foodcont.2014.03.022.

Porsbo, L.J., Jensen, T. and Nørrung, B. 2013. Occurrence and control of viruses in food handling environments and in ready-to-eat foods. pp. 181–200. *In*: Foodborne Viruses and Prions and Their Significance for Public Health, ECVPH Food Safety Assurance. Wageningen Academic Publishers. https://doi.org/10.3920/978-90-8686-780-6_09.

Potasman, I., Paz, A. and Odeh, M. 2002. Infectious outbreaks associated with bivalve shellfish consumption: a worldwide perspective. Clin. Infect. Dis. 35: 921–928. https://doi.org/10.1086/342330.

Pouillot, R., Hoelzer, K., Chen, Y. and Dennis, S.B. 2015. *Listeria monocytogenes* dose response revisited—incorporating adjustments for variability in strain virulence and host susceptibility. Risk Analysis 35: 90–108. https://doi.org/10.1111/risa.12235.

Pouillot, R., Smith, M., Van Doren, J.M., Catford, A., Holtzman, J., Calci, K.R. et al. 2021. Risk assessment of norovirus illness from consumption of raw oysters in the United States and in Canada. Risk Anal. https://doi.org/10.1111/risa.13755.

Randazzo, W. and Sánchez, G. 2020. Hepatitis A infections from food. Journal of Applied Microbiology 129: 1120–1132. https://doi.org/10.1111/jam.14727.

Rantsiou, K., Kathariou, S., Winkler, A., Skandamis, P., Saint-Cyr, M.J., Rouzeau-Szynalski, K. et al. 2018. Next generation microbiological risk assessment: opportunities of whole genome sequencing (WGS) for foodborne pathogen surveillance, source tracking and risk assessment. International Journal of Food Microbiology, Omics in MRA - the Integration of Omics in Microbiological Risk Assessment 287: 3–9. https://doi.org/10.1016/j.ijfoodmicro.2017.11.007.

Richards, G.P. 1999. Limitations of molecular biological techniques for assessing the virological safety of foods†. Journal of Food Protection 62: 691–697. https://doi.org/10.4315/0362-028X-62.6.691.

Richards, G.P. 2012. Critical review of norovirus surrogates in food safety research: rationale for considering volunteer studies. Food Environ. Virol. 4: 6–13. https://doi.org/10.1007/s12560-011-9072-7.

Rose, J.B. and Sobsey, M.D. 1993. Quantitative risk assessment for viral contamination of shellfish and coastal waters. Journal of Food Protection 56: 1043–1050. https://doi.org/10.4315/0362-028X-56.12.1043.

Ruchusatsawat, K., Nuengjamnong, C., Tawatsin, A., Thiemsing, L., Kawidam, C., Somboonna, N. et al. 2021. Quantitative risk assessments of hepatitis A virus and hepatitis e virus from raw oyster consumption. Risk Analysis n/a. https://doi.org/10.1111/risa.13832.

Rutjes, S.A., Italiaander, R., van den Berg, H.H.J.L., Lodder, W.J. and de Roda Husman, A.M. 2005. Isolation and detection of enterovirus RNA from large-volume water samples by using the NucliSens miniMAG system and real-time nucleic acid sequence-based amplification. Applied and Environmental Microbiology 71: 3734–3740. https://doi.org/10.1128/AEM.71.7.3734-3740.2005.

Rutjes, S.A., Lodder-Verschoor, F., Van der Poel, W.H., van Duynhoven, Y.T.H.P. and de Roda Husman, A.M. 2006. Detection of noroviruses in foods: a study on virus extraction procedures in foods implicated in outbreaks of human gastroenteritis. Journal of Food Protection 69: 1949–1956. https://doi.org/10.4315/0362-028X-69.8.1949.

Schwab, K.J., Neill, F.H., Estes, M.K., Metcalf, T.G. and Atmar, R.L. 1998. Distribution of norwalk virus within shellfish following bioaccumulation and subsequent depuration by detection using RT-PCR. J. Food Prot. 61: 1674–1680. https://doi.org/10.4315/0362-028X-61.12.1674.

Seymour, I.J. and Appleton, H. 2001. Foodborne viruses and fresh produce. J. Appl. Microbiol. 91: 759–773. https://doi.org/10.1046/j.1365-2672.2001.01427.x.

Sun, Y., Laird, D.T. and Shieh, Y.C. 2012. Temperature-dependent survival of hepatitis A virus during storage of contaminated onions. Applied and Environmental Microbiology 78: 4976–4983. https://doi.org/10.1128/AEM.00402-12.

Teunis, P.F.M. and Havelaar, A.H. 2000. The beta poisson dose-response model is not a single-hit model. Risk Analysis 20: 513–520. https://doi.org/10.1111/0272-4332.204048.

Teunis, P.F.M., Moe, C.L., Liu, P., Miller, S.E., Lindesmith, L., Baric, R.S., Le Pendu, J. and Calderon, R.L. 2008. Norwalk virus: How infectious is it? J. Med. Virol. 80: 1468–1476. https://doi.org/10.1002/jmv.21237.

Thebault, A., Teunis, P.F.M., Le Pendu, J., Le Guyader, F.S. and Denis, J.-B. 2013. Infectivity of GI and GII noroviruses established from oyster related outbreaks. Epidemics 5: 98–110. https://doi.org/10.1016/j.epidem.2012.12.004.

Treagus, S., Wright, C., Baker-Austin, C., Longdon, B. and Lowther, J. 2021. The foodborne transmission of hepatitis E virus to humans. Food Environ. Virol. 13: 127–145. https://doi.org/10.1007/s12560-021-09461-5.

Wang, D., Wu, Q., Kou, X., Yao, L. and Zhang, J. 2008. Distribution of norovirus in oyster tissues. Journal of Applied Microbiology 105: 1966–1972. https://doi.org/10.1111/j.1365-2672.2008.03970.x.

Ward, B.K. and Irving, L.G. 1987. Virus survival on vegetables spray-irrigated with wastewater. Water Research 21: 57–63. https://doi.org/10.1016/0043-1354(87)90099-6.

Ward, R.L., Bernstein, D.I., Young, E.C., Sherwood, J.R., Knowlton, D.R. and Schiff, G.M. 1986. Human rotavirus studies in volunteers: determination of infectious dose and serological response to infection. J. Infect. Dis. 154: 871–880. https://doi.org/10.1093/infdis/154.5.871.

Zhang, Y., Qu, S. and Xu, L. 2019. Progress in the study of virus detection methods: The possibility of alternative methods to validate virus inactivation. Biotechnology and Bioengineering 116: 2095–2102. https://doi.org/10.1002/bit.27003.

Zwietering, M.H. and Havelaar, A.H. 2006. Dose-response relationships and foodborne disease. Food Consumption and Disease Risk: Consumer-Pathogen Interactions 422–439.

Index

IIIIIIIIIIIIIIIIIIIIIIIIIIIIIIIIIIIIIIIIIIIIIIIIIIIIIIIIIIIIIIIIIIIIIIIIIIIIIIIIIIIIIIIIIIIIIIIIIIIIIIIIIIIIIIIIIIIIIIIIIIIIIIIIIIIIIIIIIIIIIIIIIIIIIIIIIIIIIIIIIIIIIIII

For Product Safety Concerns and Information please contact our EU
representative GPSR@taylorandfrancis.com
Taylor & Francis Verlag GmbH, Kaufingerstraße 24, 80331 München, Germany

www.ingramcontent.com/pod-product-compliance
Lightning Source LLC
Chambersburg PA
CBHW070716220326
41598CB00024BA/3176

9 781032 204215